International Perspectives on Mathematics Education

The International Perspectives on Mathematics Education Series

Leone Burton, *Series Editor*

Multiple Perspectives on Mathematics Teaching and Learning
  Jo Boaler, *editor*

Researching Mathematics Classrooms: A Critical Examination of Methodology
  Simon Goodchild and Lyn English, *editors*

Which Way Social Justice in Mathematics Education?
  Leone Burton, *editor*

Mathematics Education within the Postmodern

Edited by
Margaret Walshaw

INFORMATION AGE
PUBLISHING

80 Mason Street • Greenwich, Connecticut 06830 • www.infoagepub.com

Library of Congress Cataloging-in-Publication Data

Mathematics education within the postmodern / edited by Margaret Walshaw.
   p. cm. – (International perspectives on mathematics education)
Includes bibliographical references.
ISBN 1-59311-130-4 (pbk.) – ISBN 1-59311-131-2 (hardcover)
1. Mathematics–Study and teaching. I. Walshaw, Margaret. II. Series.

QA11.2.M278 2004
510'.71–dc22

                                  2004008231

Copyright © 2004 Information Age Publishing Inc.

Printed in the United States of America

CONTENTS

part III
Postmodernism within the Structures of Mathematics Education

PREFACE

Postmodern perspectives have made a significant impact on educational theory, research and policy, shaping not only the thinking of those working in education but also framing their practices. Within mathematics education a few single-authored volumes have generated interest in and discussion over postmodern thought. *Mathematics Education within the Postmodern*, as the first edited volume produced from the perspective of mathematics educational thinkers, makes an important contribution to that discussion. The papers presented in this volume are all very welcome because they challenge accepted wisdoms about both the nature of mathematics and of education. They bring to bear on their intersection a postmodern sensibility which challenges the grand narratives of mathematics education. It is a groundbreaking volume in which each of the chapters develops for mathematics education the importance of insights from mainly French intellectuals of the post: Foucault, Lacan, Lyotard, Deleuze. Having attempted to bring together Lacan, Foucault and mathematics education in the 1980s, this is not only exciting and nourishing for me personally, but it feels like support for things I tried to do some years ago, when no-one else was doing this kind of work. That kind of position can be rather lonely and to discover that this kind of work has really taken off in the most insightful of ways, with profound implications for mathematics education, is of considerable importance to me.

The chapters collected in *Mathematics Education within the Postmodern* provide insights for research, practice and policy. They address issues relevant to mathematics education, not from the discipline's familiar viewpoints, but towards theory development for mathematics education in contemporary society. What is particularly important is the way in which their analyses of the discursive practices that make up mathematics education allow us to think differently about researching and teaching mathematics. My own work stressed the central importance of the practices themselves in producing the mathematics learner as subject—a subject who could invoke not the certainty of an absolute power over nature but an omnipotent fantasy over a calculable universe, a fantasy at the heart of the production of rational economic man. This fantasy has long been one that has excluded women, who have been understood at least since the Greeks as the embodiment of the irrational and therefore the antithesis of rational man. This means that work in mathematics education which opens up the fictions, fantasies and plays of power inherent in mathematics education is important in challenging the taken for granted assumptions which underlie Western liberalism. It allows us to think about the different generation of mathematics and mathematicians and therefore of the possibility of a different world, than one thought of in terms of domination over nature/femininity.

This is an opportune time for *Mathematics Education within the Postmodern* to open up a different conceptual world. Readers will find much in the chapters to challenge their received ideas and much in the authors' evident commitments to postmodernism for reviewing those ideas. I cannot overemphasise therefore the importance of the brave and critical work presented here. It offers work that, if taken seriously, allows us to move towards no less than a different and more equitable form of social organization.

—Valerie Walkerdine
Cardiff University

FOREWORD

It is with pleasure that I introduce this latest book in the series International Perspectives on Mathematics Education. I believe that the series is recognized for striking out and charting somewhat new territory. With this book, Margaret Walshaw, as she says in Chapter 1, takes us into an area where mathematics education and postmodernism "have rarely addressed each other." The importance of opening such a dialogue is obvious. Not only do different approaches and different voices influence each other but also, only by attending and listening to the other, can researchers map where, and in what ways, they might make use of what is on offer.

And what Margaret Walshaw offers readers is "the challenge ... to understand mathematics education within the intellectual shift prompted by postmodernism while acknowledging that the modernist past continues to shape the contours of the discipline." She continues: "The postmodern sensibility in this book shifts the focus from foundations and familiar struggles of establishing authority towards exploring tentativeness and developing scepticism of those principles and methods that put a positive gloss on fundamentals and certainties." How does the book accomplish this? Like all those in the series, this book is international in its contributions, making use of the fact that settings differ, theoretical positions are far from uniform, familiarity with the postmodern approach ranges from a long-term engagement with the questions that it poses, to a comparatively recent one. However, all the chapters share a desire to disturb the taken-for-granted, to exhibit discomfort with rationality, 'truth' and certainties and to follow a rejection of the assumptions of modernity wherever it may lead.

For readers who have yet to engage with the results of thinking about mathematics education from a postmodern perspective, foreshowing a

Mathematics Education Within the Postmodern, pages ix–x
Copyright © 2004 by Information Age Publishing

modernist uniformity, the book extends a number of different alternatives. All involve adopting a discursive approach to querying power, knowledge and the strength, within mathematics education, of meta-narratives to explain. But they do so from many different perspectives, some more theoretically formulated, others embedded in classroom practices. There are authors in this book who query classroom practices by "thinking the unthinkable." Others utilise a range of theoretical positions such as, for example, psychoanalysis, that might not have occurred to readers as facilitating thinking about mathematics education. Some authors make use of the advantage offered by postmodern approaches to question such complex and by no means straightforward issues as ethics or researcher/participant roles in a context such as curriculum development.

Given that one major influence of postmodernism is the recognition of the discursive as a feature of both research and practice, engaging with the results of its application to mathematics education can only be beneficial, whether that engagement moves a reader into a use of postmodern thinking and practices, or not. It is in that spirit, therefore, that I commend this book to readers of this series.

—Leone Burton
Professor Emerita of Mathematics Education

ACKNOWLEDGMENTS

I would like to thank all the contributors to this collection for responding to the difficult task of linking mathematics education with postmodernism.

Many friends and colleagues have provided intellectual guidance along the way. In particular, I wish to thank the following people who have so generously given of their time. Their incisive readings of the chapters and thoughtful questions and comments are most appreciated.

Jack Bana	Bill Barton
Andy Begg	Pat Forster
Helga Jungwirth	Steve Lerman
Gloria Stillman	Colleen Vale

I want to give special thanks to the series editor, Leone Burton, who has offered helpful and stimulating feedback. I am also grateful to Mary Klein for her enthusiasm and for the conversations that provided the inspiration for the book.

Sandy Dittmer and Wendy Osborne, both at Massey University, New Zealand, provided all the necessary secretarial assistance with the manuscript. I am indebted to them.

—Margaret Walshaw

CHAPTER 1

INTRODUCTION

Postmodernism Meets Mathematics Education

Margaret Walshaw

ABSTRACT

The first chapter opens with a glance at the modernist enterprise and traces the beginnings of postmodern thinking. The postmodern sensibility is discussed as registering a sense of inadequacy of modernist thinking and acting to deal with our increasingly complex, plural and uncertain world. The chapter records how postmodernism shifts the focus from foundations and familiar struggles of establishing authority towards exploring tentativeness and developing scepticism. An overview of chapters in the volume is provided and future questions are posed for the discipline.

This book is about two different lines of inquiry that have rarely addressed each other: postmodernism and mathematics education. In coupling the two, this book allows us to explore mathematics education from the site of the postmodern. The concept of the "postmodern" captures those shifts in modern thought that have gained currency in Western cultures. Postmodernists ask questions about modernist thinking and acting, and propose

Mathematics Education Within the Postmodern, pages 1–11
Copyright © 2004 by Information Age Publishing
All rights of reproduction in any form reserved.

conceptual designs to investigate an increasingly complex, plural and uncertain world. As a crucial component of our complex societies, mathematics education is an evolving, self-critical enterprise—constantly on the lookout for creative solutions and new ways of looking, interpreting and explaining. It is those new ways that are of central interest in this book. The challenge for us is to understand mathematics education within the intellectual shift prompted by postmodernism while acknowledging that the modernist past continues to shape the contours of the discipline.

What do we know about the modernist enterprise and, particularly in relation to this volume, what expressions of knowledge, representation and subjectivity are immanent in it? We know that Descartes' critical quest for certainty, order and clarity was integral to the formulation of a modernist framework in the 17th century. From that time until recently most Western thinkers understood reality as characterized by an objective structure, accessed through reason by an autonomous subject. Cartesian epistemological assumptions, like these, are accompanied by beliefs about the kind of being the human is. Typically those ontologies are dualist in nature. They include such dichotomies as rational/irrational, objective/subjective, of nature/culture, and universal/particular. Taken together, these characteristically modernist beliefs about ontology and epistemology constitute a springboard for understanding knowledge, representation and subjectivity in mathematics education.

This thinking about the world dominated the Western intellectual tradition until doubts and suspicions prompted an epistemic shift. During the 1960s postmodernism became a central term in the vocabularies of a number of literary critics (Waugh, 1992). Their scepticism about subject-centered reason—anticipated earlier in the scholarly work of Nietzsche and Heidegger—registered a sense of inadequacy with the modernist destiny for paradigms of knowledge. During the early and mid-1970s other tendencies began to emerge and postmodern sensibilities spread to encompass many fields of cultural endeavor. At some point in the late 1970s, it was beginning to enter the full range of human sciences. Jacques Derrida, Julia Kristeva, Michel Foucault, among others, took up the postmodernist debate in France and in 1979 the introduction of postmodernism into the larger theoretical and cultural scene was made explicit with the publication of Jean-François Lyotard's *The Postmodern Condition* (translated into English in 1984). In this work Lyotard argued that the "grand narratives" of Western history and in particular, enlightened modernity, had broken down.

The factors that have brought about this intrusion into our lives are multiple. They include political and social crises of legitimation, and the resulting changing nature of economies and social structures in Western societies. These changes place complex and sometimes conflicting demands on us in ways we are barely able to understand or predict. Increasingly we are becoming aware of the complex construction of our work within the discipline emerging from, among other things, new forms of inclusive political tendencies, changing vocational needs, and advances in informatics and communication systems. The effects of these processes for mathematics education are unsettling, precisely because they do not map easily onto traditional understandings about the known and the knower. To the extent that these traditions are premised upon modernist thought, they do not sufficiently explain the types of agency and subjection exhibited in our disciplinary practice. In employing these conceptual understandings, some of us have come up against ambiguous meaning construction—experiencing a kind of theoretical unease with the given representations of knowledge associated with research, practice and policy for the 21st century.

For some time other disciplines within social science have engaged with the postmodernist challenges to the epistemological foundations of modernity. We are aware from a glance through a library database that tertiary education course initiatives and texts introduced many students and scholars to postmodern writings in the late 1980s. Reception to post-modern ideas has been mixed and some highly charged reactions to the insights of such thinking have been documented. On the whole, however, the convergence has been positive and flourishing—at least for other disciplines. Within mathematics education, very few analyses have been available. Seminal edited volumes written during the last decade (e.g., Bishop, Clements, Keitel, Kilpatrick, & Leung, 2003), while making important contributions to the discipline, have not tended to reflect the impact of postmodern theory. Emerging from those gaps, and recognizing that mathematics education itself is going through profound changes in terms of purposes, content, and methods, this international volume is an exploration into theory through the opportunities provided in the postmodern.

The volume opens up theoretical debate by registering the limiting effects of modernist ideas. Alongside the move away from what are regarded as classic disciplinary theories is an attempt to reconceptualize subjectivity, knowledge and representation differently. In that new terrain, there is no perfect uncontestable theory. Such theories have never existed. Given that there is no definitive explanation that will make confusion and complexity absolutely clear, this book is not about what-is-ahead and what-is-behind in theory. Rather, its central aim is to introduce conceptual tools and frameworks from postmodern thinking to help us develop an understanding about particular people and events at a specific time and place.

The postmodern sensibility in this book shifts the focus from foundations and familiar struggles of establishing authority toward exploring tentativeness and developing scepticism of those principles and methods that put a positive gloss on fundamentals and certainties. Implicit in the analyses and critiques, is a defense of the position that by embracing aspects of postmodernism, mathematics education would go some way toward undermining the discipline's essentialist and absolutist tendencies. There are other very good reasons for giving serious consideration to postmodernism, not the least of them being that in certain respects postmodern ideas converge with some of the theoretical departures that are currently dominating debate and discussion within the discipline. For example, the process of knowledge construction that is central to Michel Foucault's discourse and practice theory, resonates with recent attempts by situated theorists to model the subjectivity of learners in terms of community and participation (e.g., Boaler, 2000). Similarly Deleuze's (2001) ideas on process, emerging relationships and interconnections parallel, in some important ways, current developments in complexity science for understanding the learning of mathematics (e.g., Davis & Simmt, 2003).

What the interpretative, engaged and connected methodologies share with postmodernism is a discontent with impartial knowing, disinterested objectivity, and value neutrality. In all these methodological forms, questions of knowledge and subjectivity tend to exceed and destabilize the logic, methods and goals of the discipline. Although these versions of knowing express the sense that our inherited ideas have tended to underplay multiplicity, complexity and cultural specificities, none fully captures the postmodern analytical edge that invites a less certain space for research, pedagogy and practice. The particularly postmodern mode of dealing with meaning construction is a form of critique that acknowledges its own complicity in the analysis.

Knowledge, in postmodern thinking, is not neutral or politically innocent. Cognitive products are merely that—products constructed by cognitive agents, enmeshed in a site of knowledge production that is unavoidably political. It is Foucault to whom a large debt is owed for demonstrating how cognitive products are produced through non-neutral processes. By revealing the contingency of power, privilege and history on systems of knowledge, Foucault shows, in turn, how knowledge implies forms of social organization and constitutes individuals as thinking, feeling and acting subjects.

In order to highlight Foucault's meaning of knowledge production we need look no further than social narratives organized around female gender. Academic discourses specific to girls in Western countries trace a decades-long history that represents girls as victims of biology, mathematically disengaged and in need of compensatory measures. This knowledge about girls in mathematics filtered through into everyday practice as definitive and true. The category of girl thus produced tended to slide over its historical contingency, its competing representations and its ultimately unstable construction. When deficiencies are purported to be specific to all girls, irrespective of class, ethnicity, history and culture, the ways in which the category girl is generated by and subjected to the structural rules that govern discursive formations are inevitably obscured. A postmodern mode of exploration, on the other hand, denaturalizes the category girl, unpacking the provisional and historically specific relationship between the category and the forms of social governance that engender it.

This book is a contribution to the project of thinking about knowledge, subjectivity and representation in mathematics education. The contributors have developed their commitment to postmodern theory from different starting points. Located within a wide range of geographical, cultural and educational settings, some have been working in tertiary institutions and have been developing their postmodern theoretical investigations for some time around their practice. Others are newly attracted to postmodern sensibilities. Some of these contributors have come to postmodernism while working, and later reflecting on, the implications of a major study. Irrespective of their standpoint—critical discourse theory, Lacanian psychoanalysis, Deleuzian topologies, and Foucauldian theory—the frameworks they use and the questions they choose to prioritize have been shaped by a keen interest in the development of the academic study of mathematics education.

Precisely because the essays tap into a range of theoretical positions and are not organized around the scholarly work of a single postmodern writer, they do not add up to a unified argument. Some of the essays open with synopses of current thinking in mathematics education and use those descriptions as contrasts to other conceptual frameworks. Others draw on immediate and local conditions and use postmodern thinking as an explanatory and analytical tool. All contributors are, nevertheless, engaged in ongoing explorations that draw on postmodernism as a potential source of sophisticated analytical tools. The tools they use for their analyses include deconstruction, discourse analysis, and methods derived from psychoanalysis.

Underpinning the methodological diversity amongst the essays is a common thematic unity. They all share a commitment to establish a postmodern presence within mathematics education, even as they approach that interest in unique ways. The commitment is clear from the way each chapter disrupts and challenges many modernist assumptions common to mathematics education. In mapping out central theoretical tenets of postmodern thought and exploring their resignification for mathematics education, each essay stakes out new territory for the discipline.

In this volume the essays are arranged according to three key foci and it is these that provide a structuring organization for the book. The first focus, addressed in Part I, relates to the postmodern analysis of the limits of the modernist enterprise and what that might mean for fundamental ideas about mathematics, research practice, and ethical groundings in mathematics education. Part II introduces the second focus that relates to the part that the postmodernist suspension of truth and rationality and the self-certain subject might play in analyzing classroom practices. The third focus, explored in Part III, concerns processes and visions of practice as they are explored critically and reflexively within frameworks that are at odds with tendencies within some contemporary analyses within the discipline. The essays explore those issues through specific questions:

- How does the postmodern rejection of meta-narratives, and its stress on specificities, relate to the way we conceive mathematics?
- What are the implications for research from the postmodern deconstruction of concepts such as the transparency of representation and intellectual expertise?
- What are the grounds for the establishment of postmodern ethics in mathematics education?
- How does the postmodern emphasis on discourses and practice impact on beginning understandings of mathematics in terms of access and agency?
- What does the postmodern understanding of governance and control of the subject mean for active and agentic numeracy understandings?
- What are some of the implications, for students' learning, of the postmodern deconstruction of rational thinking processes?
- How do postmodern understandings of affect inform educators' work with developing cognition?
- What possibilities are there within the postmodern fragmentation of the unified subject for the development of identities of pre-service and in-service teachers?
- Where does the postmodern emphasis on the circulation, rather than possession, of power lead in terms of curriculum development and transformative practice?
- What curriculum lessons can we learn from the postmodern stress on the interconnections of relationships and process?
- What are the implications for the postmodernist scepticism of clarity of expression and unambiguous representation for assessment processes in schools?

The chapters in Part I introduce other ways of thinking about mathematics, research and ethical practices. In Chapter 2, Paul Ernest reveals how meta-narratives arising from the modernist project have shaken some convictions about the possibility of absolute claims to mathematical knowledge. He unpacks the epistemological assumptions and their associated ontological presuppositions that underpin the critical quest for certainty, order and clarity. He traces how those ideals fundamental to the Galilean universe and the Cartesian self-starting individual have been foundational in the development of mathematics as a systematic form of knowledge. Paul argues that it is no longer possible to believe in claims to universality, value neutrality, and objectivity of knowledge because the epistemological basis for such belief has been undermined. As a correction to those claims that have tended to hinder the construction of sophisticated forms of knowledge production, Paul posits social constructivism as a more inclusive approach to the production of mathematical knowledge.

Paola Valero, in Chapter 3, considers the theoretically complex question of research in the postmodern. She reflects on a postmodern attitude and how that attitude demands a rethinking both of the question of research authority and of ways of representation. Exploring these theoretical and methodological questions through an empirical study within a specific educational setting in Colombia, her agenda is in theorizing how research might be positioned as a productive site for pursuing the questions of emancipatory research practice. Paola considers the choices of evidence and the various representations of that evidence by the researcher. Integral to her analysis is a concern with the representation of the participants of research. She asks that we question the sorts of research identities and subject positions we write about and question, too, the conceptual needs that those identities advance or eliminate. A self-conscious consideration of the location of the researcher can highlight the processes of meaning making and consciousness, and increase our curiosity about the activities of researchers and respondents in the field.

In Chapter 4, Jim Neyland confronts the dichotomy that arises from linking the postmodernists' suspicion of truth claims to the ethical ground of mathematics education. Jim's central argument is that an orientation to ethics for postmodern times does not involve an outright dismissal of the ethical problems that guide modern thinkers but of the specifically modern approach to confronting those problems. He points out the effect on teachers' practice resulting from a loss of professional-ethical autonomy when educational reform is approached according to the principles and procedures of scientific management. Taking mathematics education as the "paradigm case subject" for arguing against the modernist project in education, Jim applies principles from the ethical philosophy of Levinas, and ideas derived by others from Levinas' work, to suggest that ethical responsibility precedes all engagement with the Other and that such engagement is not dependent on the reciprocation of the Other. Jim makes clear that he is not offering a panacea through a postmodern approach to ethics but is suggesting that such an engagement will be more productive and creative because the focus will shift from teacher compliance to direct ethical teacher-student relationships.

In Part II, the authors explore parameters and possibilities for a postmodern analysis of classroom practices. Agnes Macmillan, in Chapter 5, argues that children's immersion in social contexts is fundamental to the development of numerical literacy. She contends that sociocultural processes provide the necessary resources for the development of a consciousness of abstract meanings and an appreciation of their symbolic potential. In a researcher-as-teacher investigation set in Australia, Agnes explores how children use particular forms of spoken and written language to develop an understanding of basic mathematics. She draws on the postmodern understanding of the discursive production of meaning and uses the tools of critical discourse analysis to develop her analysis. She maintains that certain socioregulative relations are at play in the classroom: on the one hand, the agentic learner is spontaneous, curious and creative, and, on the other, the teacher holds certain authorized meanings. Agnes notes that irrespective of the teacher's desire to accept children's idiosyncratic interpretations, simultaneously, a desire to protect the meanings and conventions of mathematics is crucial to pedagogical practice. She discusses how this dilemma is resolved.

In Chapter 6, Tansy Hardy makes use of Foucault's notion of normalization to unpack hidden relationships and practices of governance operating within the classroom. She investigates those aspects in relation to a professional development classroom resource distributed to teachers of numeracy within the United Kingdom. Through the video resource Tansy demonstrates the ways in which the discursive practices of mathematics education position people and contribute to the development of thinking in the classroom. Discursive practices shape thinking by limiting the scope of what can be said and done. Tansy is concerned that issues surrounding the producers of knowledge need to be addressed and asks that teachers critically assess nationally distributed "exemplary" teaching and learning practices. Applying to her investigation the postmodern perspectival mode toward representation, Tansy marks out her own invested observations with the research.

In Chapter 7, I consider the rational learner of mathematics education and traverse other boundaries of knowing. After tracking the configurations of the learner that have currency in two theories of learning, I attempt to formulate a viable and credible analysis of student learning, when confronted with learning activity in a research project that appeared to exceed and transgress available discourses. To account for the unexplained and the unsaid, I apply a theoretical approach whose precepts are in tenor unsympathetic with the presuppositions that underwrite the rational autonomous learner. Ideas formed at the conjunction of Foucauldian poststructuralism and Lacanian psychoanalysis are drawn upon in order to develop a conception of the learner who is an historically particular, social, embodied, and interested individual, at once both rational and irrational. What emerges in the analysis is a foregrounding of the place of power and the unconscious, and a focus on the play of affect and how it attaches itself in the desire to learn.

In Chapter 8, Tânia Cabral draws on her teaching experiences in a psychoanalysis-inspired learning environment in Brazil to discuss an approach to resolving learning difficulties and misconceptions. Her central theme is affect and anxiety. In first outlining the work on affect undertaken in mathematics education, she argues that research to date is premised on the conscious self and, hence, overlooks important processes that take place in the unconscious. She outlines Lacan's four fundamental concepts of psychoanalysis and, from these, proposes the idea of pedagogical transference. Crucial to pedagogical transference and to knowledge production is language. Tânia shows how the concept of pedagogical transference is linked to student motivation and thus how the teacher's desire might be connected to transference. In applying these psychoanalytic ideas to learning, Tânia offers vignettes of two integrated sessions in which a number of educators assist the mathematics learning of two tertiary level students.

Part III investigates what the postmodern understanding of the limits of human knowledge and agency might mean for structures and processes in mathematics education. Tony Brown, Liz Jones, and Tamara Bibby, in Chapter 9, consider identity from a postmodern perspective and trace the changeable and unpredictable identity constructions of first year students and first year teachers. In outlining the processes of the self that are undertaken in the transition as students move from being learners to teachers of mathematics, the authors capture a reconciliation between constructions of past, present and future possible identifications. In particular, they explore how constructions made available in government initiatives compete for the individual teacher's attention and both summon and dismiss earlier identifications. Importantly, the authors do not slide over the heavy historical emotional baggage that some teachers identify with mathematics. They link that history to constructions of mathematics and sketch out the ways in which mathematics is emergent through the teachers' appropriation of various social discourses of mathematics teaching. They suggest how teachers' multiple constructions of mathematics might be productive in the professional development of teaching.

Chapter 10 outlines a critical approach to a curriculum development project. Tamsin Meaney describes her attempts as curriculum consultant to provide an empowering and negotiated approach to curriculum development in a small indigenous school community in New Zealand. Confronted with community actions and decisions that tended to perpetuate relations of domination between "outside expert" and community, Tamsin looked beyond the Freirian model of empowerment to try to understand the particular relationship between her role as agent of empowerment and those who were to be empowered. She provides a model of her behaviors as consultant and through an innovative treatment of Foucault's power-knowledge couplet she explores how power circulates in a productive network through her relationships with the Māori immersion school community. Tamsin's project raises questions about action and curriculum change and intimates that empowering approaches to curriculum development should be used more cautiously and reflexively in the literature.

Chapter 11 explores curriculum through spaces made available by Deleuzian (2001) insights about process and relationships. M. Jayne Fleener challenges us to reflect on the constraints of our current thinking, noting how exceptional work undertaken in mathematics has often developed from those who dared to think differently. Jayne develops the theme of curriculum interconnections by drawing on the analysis of teaching as provided by Roy (2003). She introduces key words that are integral to Roy's analysis—rhizome, in-betweeness, nexus and becoming—and applies them to curriculum work in the mathematics classroom. On the basis that over-prescriptive curricula, control and planning tend to impose limits on creativity and self-organization, she advocates that learning environments might begin by nurturing multiplicity and difference and might consider supporting openness and exchange of energies.

In Chapter 12, Tony Cotton investigates what the "crisis of representation" might mean for mathematics education research and applies that investigation to the issue of assessment. Tony draws on Lyotard to argue that the problem is not research itself, but the presentation of prescriptions and supposed fixities emanating from research. In undermining traditional understandings of scientific method, he suggests that researchers need to keep foremost in mind that reality does not have an objective structure, that research is fundamentally unrepresentable, and that representation is subjective and hence highly contested. Adopting a Foucauldian approach, and drawing on school-based research, Tony presents an archaeology of assessment processes relevant to primary/elementary schools in the United Kingdom. Underlying that presentation is an intent to avoid mere descriptions and to center the analysis on why things are as they are and how they might be different. He sketches out models of assessment and offers an alternative practice based on authenticity and high levels of participation of students.

For all the authors, mathematics is a central component of the complex societies in which they live. They demonstrate courage in studying mathematics education not in isolation but head-on as a disciplinary endeavor situated at the interface of multiple and competing structures and processes. In advocating engaged forms of critique that acknowledges their own complicity with particular educational principles, values and emancipatory potential, their interrogations go some way to refute the relativism with which postmodernism is often labeled.

Arguably, the volume opens up theoretical discussion for the discipline. Implicit is an invitation to the reader to explore how postmodern theoretical frameworks raise, and provide possible responses to, certain issues for the discipline. What such exploration would reveal is that theoretical border crossing into the postmodern does not necessarily entail a wholesale rejection of conventional disciplinary ideas. What it does require is a shift in attitude. It also demands our attention to our ultimately compromised stance in everything we do and say.

Further exploration will move the discussion and will spark other questions. For example, we might ask: What postmodern power-knowledge lessons might be learnt for the discipline from the recent reconstruction of academic identities and new work environments, centered as they are on performativity and measurable research and publication outcomes? Where does the postmodern collapse of the distinction between knowledge and commodity, with regard to technology, lead to in terms of the production of mathematical knowledge? Responses to these and other questions, as yet unarticulated, will keep the conversation going, encouraging further engagement with, and further insightful analyses from interrogations made possible through the postmodern. Ultimately it is the hope of all the authors that the ongoing engagement will mark a fruitful and productive convergence between mathematics education and postmodernism.

REFERENCES

Bishop, A., Clements, M.A., Keitel, C., Kilpatrick, J., & Leung, F. (Eds.) (2003). *Second international handbook of mathematics education.* Dordrecht: Kluwer Academic Publishers.

Boaler, J. (Ed.) (2000). *Multiple perspectives on mathematics teaching and learning.* Westport, CT: Ablex Publishing.

Davis, B., & Simmt, E. (2003). Understanding learning systems: Mathematics education and complexity science. *Journal for Research in Mathematics Education, 34*(2), 137–167.

Deleuze, G. (2001). *Logic of sense* (Trans: M. Lester with C. Stivale; Ed: C.V. Boundas). London: Athlone.

Roy, K. (2003). *Teachers in nomadic spaces: Deleuze and the curriculum.* New York: Peter Lang.

Waugh, P. (Ed.) (1992). *Postmodernism: A reader.* London: Edward Arnold.

part I

THINKING OTHERWISE
FOR MATHEMATICS EDUCATION

Part I relates to the postmodern critique of the modernist enterprise and considers what that critique might mean for fundamental ideas within mathematics education. The three chapters in Part I consider the implications of the postmodernist shift from conceptual absolutes to contingencies. They introduce, in turn, other ways of thinking about mathematics, research practice and ethical groundings.

CHAPTER 2

POSTMODERNISM AND THE SUBJECT OF MATHEMATICS

Paul Ernest

ABSTRACT

This chapter looks beyond modernist visions and foundations towards post-modern views of knowledge production which foreground human language, culture, community, conversation and history. Lyotard, Wittgenstein, Derrida, among others, are drawn upon to link postmodern ideas with Social Constructivist perspectives of mathematics education. The linkage problematizes mathematics and school mathematics, and suggests ways of reconceptualizing classroom knowledge. Finally it is suggested that the postmodern conceptualization of multiple selves is a potentially rich area for investigation.

INTRODUCTION: THE META-NARRATIVE OF MODERNISM

When Descartes (1637) ushered in modernism with his logical master plan, he hoped to provide certain and indubitable foundations for all knowledge. Beginning with a small basis of clear and distinct ideas, in Descartes' formulation, all truths are subsequently deduced, modeled on the axiomatic geometry of Euclid. Descartes' aim was that rationality would

Mathematics Education Within the Postmodern, pages 15–33

illuminate all the darkness of ignorance, sweeping away traditional superstition and reliance on the ancients. In his vision, human language and culture, community, conversation and historical agreements within groups of knowers, are all irrelevant contingencies in the genesis, warranting and status of knowledge. The implications of the modernist worldview, for all disciplinary knowledge, are profound, and what we have witnessed for some time now in mathematics education is an opposition to the exclusions it establishes (see Bishop, 1988; Bloor, 1991; Ernest, 1991; Restivo, 1992; Restivo, Van Bendegen, & Fischer, 1993; Skovsmose, 1994).

Early twentieth century philosophy of mathematics applied Descartes' model to mathematics itself in the quest for absolute foundations for mathematical knowledge. The logicist, intuitionist and formalist philosophies all sought, each in their characteristic ways, to devise narratives of certainty. Each started with "self-evident" truths, proceeded to offer a safe and certain foundation for mathematical knowledge, and then adopted what were hoped to be guaranteed means to extend the certainty throughout the field. The legacy of this project can be gleaned from popular understandings of mathematics as an unquestionable certain body of knowledge. However these beliefs and personal theories run counter to arguments (e.g., Davis & Hersh, 1980; Hersh, 1997; Kitcher, 1984; Kline, 1980; Lakatos, 1976; Tiles, 1991; Tymoczko, 1986) about the limitations of Descartes' modernist vision for mathematical knowledge.

This chapter looks beyond the modernist vision for mathematical knowledge toward postmodern views of knowledge production. Central to the discussion is the part that human language and culture, community, conversation and history all play in what we mean by mathematics. To my way of thinking, Descartes' ideas about logic and rationality necessarily rest on narratives, which shift with changes of culture over time. I make a case that mathematics—its practices and its theories—like all human knowledge and, indeed, like postmodernism itself, avoids absolute characterization. It is, rather, an outflow of conversation. In that understanding, mathematics consists of language games with deeply entrenched rules and patterns that are very stable and enduring, but which always remain open to the possibility of change, and, indeed, in the long term, do change.

To support my argument I introduce the work of a number of postmodern writers (Lyotard, Wittgenstein, Derrida, and others) and apply a range of ideas from their work to a discussion about the production of mathematical knowledge. I draw connections between these postmodern ideas and social constructivist perspectives of mathematics education in order to suggest how we might conceptualize classroom knowledge. Finally I point to a potentially rich area for investigation, one that takes into consideration the postmodern conception of multiple selves.

POSTMODERN MATHEMATICS

Lyotard (1984), like other postmodern writers, questions the modernist secure basis of knowledge. Traditional objective criteria of knowledge and truth within the disciplines, according to him, are nothing but myths that deny the social basis of knowing. Lyotard maintains that knowledge consists of narratives and each disciplined narrative has its own legitimation criteria. These are internal to the discipline and develop to overcome or engulf contradictions. Lyotard describes how, in this way, mathematics overcame the crises in the foundations of axiomatics brought about by Gödel's (1931) Theorem. He claims that meta-mathematics were incorporated into its enlarged research paradigm. He also notes that continuous differentiable functions are losing their preeminence as paradigms of knowledge and prediction, as mathematics incorporates undecidability, incompleteness, Catastrophe theory and chaos. Thus, a static system of logic and rationality does not underpin mathematics, or any discipline.

Lyotard's perspective, like a number of other intellectual traditions, asserts that all human knowledge is interconnected through a shared cultural substratum, and is a social construction. Certainly the history of mathematics bears out Lyotard's reading. Mathematics has been defusing uncertainty by colonization since its beginnings. This includes the incommensurability of lengths (i.e., the irrationality of the square root of two), Zeno's paradoxes, the Delian problems, negative numbers, probability, infinitesimal numbers, transcendental numbers, statistics, Cantor's set theory, infinity, Peano's and other space filling curves, logical paradoxes, computer proofs (e.g., the four color theorem), as well as those uncertainties that Lyotard mentions. Each of these topics has caused great philosophical anxiety and controversy. But despite the temporary anxiety, all these topics are widely accepted as technical and conceptual advances, and ultimately not as challenges to the underlying paradigm of rational control and scientific certainty. Gödel's Theorem did not even cause mathematics to break its stride as it stepped over this and other limitative results. Chaos is merely the latest branch of mathematics to be tamed and engulfed, and not the beginning of a wholly new paradigm. Like modern art when responding to the challenges that Impressionism, Cubism, Dada, Surrealism, Abstract Expressionism, Pop Art, and Brit Art presented to its boundaries, mathematics continues to advance and churn out new knowledge.

It would seem that secure foundations for mathematics are constantly being undermined (Davis & Hersh, 1980; Kitcher, 1984; Kline, 1980). As far as Wittgenstein is concerned, mathematical foundations are quite irrelevant to the continued healthy practice of mathematics. Through his central concepts of language games and forms of life, Wittgenstein (1953, 1978) offers a powerful social vision of mathematics. He argues that following a rule in mathematics or logic does not involve logical compulsion. Instead it is based on the tacit or conscious decision to accept the rules of a "language game" that are grounded in pre-existing social "forms of life." Wittgenstein shows that the "certainty" and "necessity" of mathematics are the result of social processes of development, and that all knowledge, including that in education, presupposes the acquisition of language in meaningful, already existing, social contexts and interactions. That means that no particular area of mathematics has a fixed and unique mathematical theory formulation as source since there are always multiple formulations provided by different mathematicians and groups of mathematicians. Unlike Foucault, however, Wittgenstein fails to clarify the historical development of social structures and linguistics use-patterns. Thus he remains trapped in the timeless, tenseless discourse of academic Anglo-American philosophy.

Foucault (1972) argues that the divisions of knowledge accepted today are modern constructs, defined from certain specific social perspectives. The objects, concepts and aims of disciplinary knowledge, and the accepted rules for formulating and validating that knowledge, have all evolved and changed. As evidence, he documents how certain socially privileged groups, such as doctors and lawyers in the last century established discourses which created new objects of thought, grouping together hitherto unconnected phenomena defined as, for example, delinquent behavior or crime. In one study he shows how a new subject area, the discourse of human sexuality, was developed by church and state, for their own reasons (Foucault, 1981).

Lakatos (1976, 1978) traces the development of mathematical knowledge as evolving and changing. He argues that the historical and conceptual change-basis of the concepts, terms, symbolism, theorems, proofs, and theories of mathematics are central to its philosophy. Lakatos reveals how mathematical knowledge is contingent on historical accident, as well as driven by internal problems and principles. He demonstrates that the methodology of mathematics, as used by practicing mathematicians, does not differ in kind from the heuristics of problem solving and knowledge production in the classroom. Both are social processes. Lakatos shows the historical and present-day import of conventions, agreement and power in the warranting of mathematical knowledge.

Just as there is no essence to mathematics itself, so too there is no essence to school mathematics or to mathematical research topics. Boundaries change as elements from one division of school mathematics are absorbed and reconstituted into another, for different contexts and different periods of time. There is partial overlap between different topics—some are learned in parallel, and others develop and extend topics met earlier in study. The name may remain the same, but this is convention and convenience. The underlying processes and entities are all the while shifting and changing.

Lakatos (1978) and Kitcher (1984) both identify power and authority as playing a central role in communicating mathematical knowledge at both the disciplinary and the educational levels. Powerful professional groups control access to truth and knowledge through delegitimating and pathologizing alternative narratives. This process can be observed in, for example, the approach to (social) constructivism in the Science and Math Wars, and in interpretative research in mathematics education. Things change, and research is no longer wholly dominated by the meta-narrative of positivism, although the planning and governance of education still remain under its sway. As Cooper (1985) explains:

> Subjects will be regarded not as monoliths, that is as groups of individuals sharing a consensus both on cognitive norms and on perceived interests, but rather as constantly shifting coalitions of individuals and variously sized groups whose members may have, at any specific moment, different and possibly conflicting missions and interests. These groups may, nevertheless, in some arenas, all successfully claim allegiance to a common name, such as "mathematics." (p. 10)

Following Foucault and Lakatos, we can reinterpret the division and categorization of mathematical knowledge into discrete areas of study. We can question the artificial and enforced separation of applied mathematics, university research mathematics, school mathematics, ethnomathematics, accountancy, and so forth, at the same time as we can question whether "mathematics" names any unified and identifiable central area of knowledge or social practice.

QUESTIONING THE HUMAN SUBJECT

We could not learn, understand or use mathematics or be involved in the production of mathematical knowledge, unless we were social beings with personal histories and mathematical learning trajectories. Mathematical knowledge requires a mathematical knower, and this must be a fleshy, embodied human being with both a developmental history (including an educational history) and a social presence and location. Obvious as these statements are, they have been ruled irrelevant and inadmissible by generations of mathematicians and philosophers, caught up in the meta-narratives of modernism.

Central to modernism is the model of the self as essential, unitary and rational. This thinking and perceiving self inhabit the body like a "ghost in the machine" (Ryle, 1949). Within a century of Descartes' seminal contribution, David Hume (1739) had already rejected the existence of a coherent self-identical and essential self in favor of a stream of mental impressions and events. Freud theorized a multiple self incorporating three domains: Ego, Superego and Id. The Ego is the agentic wilful self, but is covertly controlled by the Superego with its rules and inhibitions, while acting out the libidinous drives encoded in the Id. This "unconscious is structured like a language" (Lacan, 1977, p. 203) because of the way it makes signifiers of objects in the world and dreams, and attaches meanings and sublimated desires to them using displacement (metonymy) and condensation (metaphor). Apparently the self is not so unified and not so rational.

Although the self is not completely unified and rational, we all share, Wittgenstein argues, some indescribable underlying reality, or at least a part of one. In some pre-philosophical sense human beings all have the experience of living together on the Earth. As a common species we have comparable bodily functions and experiences that make commensurable our sense of being who we are. Heidegger's (1962) view is that we all have a given, "thrown" preconceptualized experience of being an embodied person living in some society. He celebrates our "being-here-now" existence (*Dasein*), acknowledging our multiple existence in the linked but disparate worlds of our experience, the bodily, discursive, political, and cultural realms. This experience is taken for granted, and provides the grounds on which all knowing and philosophy begins, although no essential knowledge or interpretation of the basal lived reality is either assumed or possible. This is a postmodern bottom-up perspective that contrasts with the top-down position of modernism in which the "gaze" of a reasoning Cartesian subject, with its legitimating rational discourse, precedes all knowledge and philosophy. The rational knower does not come first. He (and I use the masculine deliberately) is not a universal disembodied intelligence, but a construction with historically shaped sensibilities.

Postmodernism rejects the fiction of a Cartesian self in favor of a multiply fragmented set of selves-in-context. The social view builds on the work of Mead (1934) and Vygotsky (1978) and sees mind-in-society at work (Lave & Wenger, 1991). Wenger (1998) theorizes the development of personal identity in terms of experiences in social practices, and since individuals are active in several social practices this involves multiple aspects of identity. Poststructuralists see self distributed over a number of different discursive practices (Foucault, 1972; Henriques et al., 1984). Persons are viewed as having multiple selves that are elicited and evidenced in different social contexts. Although a fleshy human being underpins these selves, distinct identities are constructed in different discursive practices by the different positioning of the individual through the linguistic and social arrangements in place.

From this perspective, students in mathematics classes are not identical to those in other lessons, in sport, pastimes, or in the home or on the street. Simultaneously, teachers are different in the classroom, staff meetings, and at home with their families. All of us may be very different with those superior to us in social hierarchies than we are to those positioned less powerfully. In theorizing these differences, postmodernists reveal a commitment to the irreducible multiplicity of both practices and perspectives. For them, there is no *a priori* reason to accept that even those subjects who are conventionally labeled the same way, for example, learner, girl, mathematics, classrooms, research project, school mathematical topic, are essentially the same within the category, under the stone of the name. It is rather that the observed shared characteristics, the emergent family resemblance, to use Wittgenstein's (1953) metaphor, define the term in each of these cases. And within any family there may be many differences, shifts with age, new arrivals, and deaths.

In Ernest (1989) I explored the attitudes of primary school teachers toward mathematics. I found teachers in my sample exhibited two contrasting attitudes—typically a negative attitude to mathematics itself but a positive attitude to its teaching. Being questioned about mathematics seems to have elicited memories of being positioned as struggling learners of mathematics, perhaps not too happily, given the negative self reports. In this discourse the teachers as students would have been searching without sufficient success for certainties in a discourse beyond their control, with agendas set by others. When asked about their attitudes to teaching mathematics I received positive responses. Presumably the questions tap into their positioning in the discourse and practice of teaching mathematics, where they are powerful, authoritative and setting the agenda themselves. The fact that as a researcher I regarded both discourses as concerning mathematics is irrelevant. According to my interpretation the significant factor is the different selves constructed through their positioning in the two discourses. The "mathematics teacher" self is autonomous, powerful and in control and, not surprisingly, has positive attitudes. The "mathematics learner" self is weak, subjected and not in control, and again, not surprisingly, has negative attitudes to mathematics.

Although my conclusions are tentative, this example indicates how positioning and different selves can be powerful ideas in educational research. One and the same individual can manifest different selves in different discursive practices according to that individual's positioning. However, the multiple, postmodern view of persons does not repudiate the commonalities that individuals carry with them into new social practices. These include the person's underlying psychological make up—not in some unalterable essential sense—but as a formed but evolving psyche with a still open history of emotional sensitivities and responses. A person also takes with her knowledge of signifiers and some of the associated skills. However, despite some "transportability," different significations, meanings and emotional responses are also activated in the second social practice, as Evans (2000) shows in his study of adults and mathematics.

LOGOCENTRISM, CONVERSATION, AND MATHEMATICAL WRITING

In Derrida's postmodern analyses signs, text and semiotics are key terms in his understanding of representation. In this section I introduce some of Derrida's ideas concerning representation because they are paramount to any discussion about reading and conversation. Derrida's conception of representation is one that rejects the modernist assumption of a unique or privileged meaning. He terms this myth Logocentrism derived from the arguably unreasonable privileging of the spoken word. According to his critique, throughout western philosophy the word is presumed to be imbued with a fixed meaning due to the presence of its speaker, the utterer. Even written texts are permeated with this myth, which in effect detaches written texts from their contexts and imbues them with some essential meaning and content stemming from their origin. Derrida claims that all writing is multiply coded, and that readings and meanings are actively constructed by readers and communities. When a published text is released to the public the readings that ensue will be the product of the readers and the particular circumstances and cultural contexts in which the text is read. There is no way of fixing unique readings, and this is as true for philosophy, science and mathematics, as it is for fiction (Anderson et al., 1986).

Derrida extends his rejection of essential word and concept to include subjects, both human individuals and disciplines/school subjects. He challenges the idea that they possess some enduring unified and defining set of properties that may be characterized as their "essential nature." For him, essentialism fails because of inescapable diversity of subjects on both the synchronic and diachronic planes. The synchronic plane, on the one hand, concerns the subject as it is presented structurally in the timeless present. This aspect is characterized by multiple and contradictory elements, whether they are different social practices or different elements of the subject as self. The coexistence of contradictory elements, like the opposite and repulsively charged particles held together by the strong nuclear force in the atomic nucleus, is part of what Derrida (1976) terms deconstruction. The diachronic plane, on the other hand, is characterized by multiple and distinct elements with trajectories that emerge and vary over time, propelled by complex interactions and historical contingencies, and not driven purely by some internal or even evolutionary logic. Thus persons, disciplines and school subjects, such as mathematics, all grow and change as inner forces and developments and external circumstances intertwine and cross over the shifting boundaries of the subject, flowing one way and ebbing the other.

Many binary oppositions cease to maintain their clear cut or essential differences under Derrida's gaze. One of the most difficult to accommodate in academic writing is the erosion of the signifier-signified, form-content dichotomies. This follows directly from Derrida's version of postmodernism in which he asserts that there is no *essential* difference within such binary pairs, although differences do exist. Slipping back and forth across the signifier-signified and form-content boundaries is manifested in Derrida's playful writing. It certainly facilitates the openness of the text, encouraging the reader to construct her own meanings.

In this understanding, the traditional academic style of reporting, arguing, discussing, expounding, with careful arguments and evidence, is challenged. In the production of the text the author operates in a semiotic space topologically distinct from the final sequencing of the text. She drafts, redrafts, reorders the body of the text to produce meanings that only become clear as the whole text emerges. It is through the process of constant redrafting and editing that mathematical texts take on their finished and polished mono-logical form. Writing mathematical texts typically inverts the order of creation (Lakatos, 1976), and the order of creation is itself far from any sequence of spoken words. The writing is not a simple coding of the spoken word, for it relies on icons, symbols, ideograms, diagrams, and so on, spread out two dimensionally over the page. Reading, too, is not a simple coding practice: one's gaze darts from focal point to point, up and down, side to side, stringing together a necklace of meanings into a larger pattern of interpretation.

Semiotic thinkers about mathematics have made this point forcibly. Rotman (1994) writes of "the distorting and reductive effects of the subordination of graphics to phonetics" (p. 76). He argues that even within mathematics, ideogrammatic symbols have been admitted as honorary literals at the expense of mathematical diagrams, graphs, figures, charts, and so on. These have been relegated as inessential explanatory or motivational addenda to the "real" (i.e., proper) alphabetic code of mathematics. But this is to do violence to how mathematics is practiced, whether in the contexts of school, research, or applications. It is another illustration of logocentrism, which is not only alive and well, but also remains dominant. Rotman terms the subsumption of all writing under the alphabet, the *alphabetic dogma*, which in the special case of mathematical writing denies the special place of both ideograms and diagrams (Rotman, 2000).

Lemke (1998, 2003), in his analysis of mathematical texts, defines the topological mode of representation as concerning continuous relations and differences in degree, typified by diagrams. He insists that the semiotics of mathematics necessitate a combination of both this and the typological (alphabetic) mode of signification, and the elimination of one in favor of the other is not only restrictive and unproductive, it renders continued progress in the field impossible. The conclusion of these and other scholars (e.g., Peirce, 1931–58) is clear: mathematical text cannot be reduced to purely alphabetics.

None of this invalidates writing-as-conversation as a central part of the epistemology of mathematics. But it does mean that the differences between live face-to-face conversation, written, and intra-personal forms must be acknowledged. A finished text can move between social practices and be inhabited by new potential meanings, whereas live conversation is parochial, limited to its place of utterance. Mathematical texts cannot be records of live or spoken conversation, and cannot be chased back to this point of origin for their modes of signification and construction are irrevocably tied into their textual or semiotic texture. Mathematical texts transcend alphabetics, and transcend the here-and-nowness of live conversation. Nevertheless, mathematics is conversational in the broader cultural sense—it is always textual, and always social, requiring a writer and presupposing a reader, who, even when it is most formally coded, must draw on its cultural meanings to make sense of it, and for it to fulfil its function.

Of course this loosening does not mean that "anything goes" neither for the writer nor the reader. Nor does it mean that rational academic argumentation is excluded. But it does mean letting go of the rationalist meta-narrative as the sole arbiter of acceptable academic writing. As author, I can only do this in part, and perhaps only superficially, having entered the academy via a restrictive group of disciplines, the physical sciences, mathematics, logic, and Anglo-American philosophy. Reacting against these past strictures may be the source of my energy and impetus to challenge the meta-narrative of absolutism in the philosophy of mathematics and mathematics education. But my "mind forged manacles" (Blake, 1794, p. 46) will always remain inscribed in my being.

SOCIAL CONSTRUCTIVISM

From the understandings put forth in the preceding sections we can interpret mathematics not as a fixed unity, but as a set of diverse multi-centered socially situated practices across time, space and institutional locations, making use of different textual forms to embody mathematical knowledge.

Contrary to Platonic and Popperian (1970) myths, mathematical knowledge, like everything else, is grounded in the material world, and in particular, is located in historical and discursive practices, produced by mathematicians, teachers and learners, for whom language is a fundamental organizing principle. These understandings are central to social constructivism.

Social constructivism (Ernest, 1991, 1998; Hersh, 1997; Restivo, 1992; Restivo et al., 1993) claims that the concepts, definitions, and rules of mathematics (including rules of truth and proof) were invented and evolved over millennia. Thus mathematical knowledge is based on contingency, due to its historical development and the inevitable impact of external forces on the resourcing and direction of mathematics. Much of mathematics follows by logical necessity from its assumptions and adopted rules of reasoning, just as moves do in the game of chess. Once a set of axioms and rules has been chosen (e.g., Peano's axioms or those of group theory), many unexpected results await the research mathematician. This does not contradict the skeptical epistemology of social constructivism for none of the rules of reasoning and logic in mathematics are themselves absolute.

Just as the historical construction of mathematics and mathematical knowledge is central to social constructivism, so too is the social aspect of knowledge. Knowledge production is based on the deliberate choices and endeavors of mathematicians, elaborated through extensive processes of reasoning. Since both contingencies and choices are at work in the creation of mathematics, it cannot be claimed that the overall development is either necessary or arbitrary. Following Bloor (1984), Harding (1986), and others, mathematical knowledge is understood as social, cultural, and public, and not as external, absolute or otherwise extra-human. Mathematics is viewed as basically linguistic, textual and semiotic, but embedded in the social world of human interaction. The form in which this is embodied in practice is in conversation, understood in the extended sense of Rorty (1979), Harré (1983), Shotter (1993), and many others who take conversation as a basic epistemological form.

> If, however, we think of "rational certainty" as a matter of victory in argument rather than of relation to an object known, we shall look toward our interlocutors rather than to our faculties for the explanation of the phenomenon. If we think of our certainty about the Pythagorean Theorem as our confidence, based on experience with arguments on such matters, that nobody will find an objection to the premises from which we infer it, then we shall not seek to explain it by the relation of reason to triangularity. Our certainty will be a matter of conversation between persons, rather than an interaction with non-human reality. (Rorty, 1979, pp. 156–157)

The original form of conversation is that of persons exchanging speech, based on shared experiences, understandings, values, respect, and so on. That is, it consists of language-games situated in human forms of life. Two secondary forms of conversation are derived from this. Intra-personal conversation is thought as constituted and formed by internalized conversation with an imagined other (Vygotsky, 1978; Mead, 1934). There is also cultural conversation, which is an extended version, consisting of the creation and exchange of texts in materially represented form beyond the transitory patterns of live speech. The extended cultural form of conversation introduces a major shift in representational possibilities.

These three forms of conversation are social in manifestation (interpersonal and cultural), or in origin (intra-personal). Conversation has an underlying dialogical form of ebb and flow, comprising the alternation of voices in assertion and counter assertion. Conversation is the source of feedback, in the form of acceptance, elaboration, reaction, criticism and correction essential for all human knowledge and learning. The different conversational roles involved include two particularly notable voices or modes. First, there is the role of proponent or friendly listener following a line of thinking or a thought experiment sympathetically, for understanding (Peirce, 1931–58; Rotman, 1993). Second there is the role of critic, in which an argument is examined for weaknesses and flaws. Both roles are necessary in fruitful conversation, at any level.

Taking conversation as an epistemological starting point has the effect of re-grounding mathematical knowledge in physically-embodied, socially-situated acts of human knowing and communication. It rejects the Cartesian dualism of mind versus body, and knowledge versus the world. It acknowledges that there are multiple valid voices and perspectives on knowledge. And, as Habermas (1981) notes, this acknowledgment also has significant ethical implications.

There are specific features of mathematics that support the claim that mathematical knowledge and knowing are conversational:

1. Mathematics is primarily a symbolic activity, using written inscription and language to create, record and justify its knowledge (Rotman, 1993, 2000). Viewed semiotically as comprising texts, mathematics is inescapably conversational for it must always address a reader. As Volosinov (1973) says, "In all cases *the word is orientated towards an addressee*" (p. 85) The reading of any text is dialogical, with the reader interrogating it and creating answers from it.

2. Many mathematical concepts can be analyzed to reveal dominant underlying meanings or interpretations that are at root dialogical and conversational (Ernest, 1994a).

3. Mathematical proof, so central to mathematical epistemology, originally developed from a cultural practice of disputation, i.e., conversation, and several modern developments in proof theory, still treat proofs as part of a dialogue.

4. The acceptance of mathematical knowledge depends on a social mechanism that mirrors the structure of conversation. We might call this the Generalized Logic of Mathematical Discovery (Ernest, 1998).

However, a word of caution is needed. Although mathematics is claimed to be at root conversational, it is also the discipline, more than any other, that hides its dialogical nature under its monological appearance, and has hidden the traces of multiple voices and of human authorship behind a rhetoric of objectivity and impersonality. This is why the claim that mathematics is conversational might seem so surprising. It rejects the traditional modernist view of mathematics as disembodied, superhuman, monolithic, certain and eternally true.

Social constructivism links the learning of mathematics and research in mathematics in an overall scheme in which knowledge travels either embodied in a person's capabilities or in a text. The social processes of formation and warranting in the two contexts are analogous. Both involve apprenticeship and participation in a productive practice. In each case conversation has a shaping role in the emergence of texts and identities.

During the process of mathematical activity, mathematicians and others engage in extended work with texts and symbols and consequently construct such powerful internalized meaning structures and convincing imagined "math-worlds" in which the objects (signifieds) of mathematics seem to have an independent existence and reality (the plausibility basis of Platonism). Educationally, this parallels an inverted problem that symbolic manipulations often do not lead to the construction of subjective "math-worlds," and this leads to problems of incomprehension, alienation and failure. The situation invites a paradox in mathematics education: mathematics is very clear and reasonable, yet when the reasoning is not understood it becomes the most irrational and authoritarian of subjects.

Social constructivism welds together the different practices of research mathematics and the learning of mathematics. Mathematicians' capabilities and identities are irrevocably shaped by studying (and teaching) it. Locating mathematics in embodied real world practices implies that mathematicians' identities, meaning making and semiotic capabilities are necessarily relevant to mathematics and its philosophy. As human beings mathematicians operate in a cultural realm, not some disembodied space of pure ideas. Every human being starts as a child and gradually acquires the linguistic and semiotic basis for doing mathematics. When children begin schooling they can already sort, count, locate, play, make, design, plan, explain, argue, and maybe measure. These are the activities that Bishop (1988) identifies as the cultural basis of mathematics. By building on this foundation future citizens (including mathematicians) are socially constructed.

IMPLICATIONS FOR CLASSROOM WORK

If mathematics is not a fixed unity, but a set of diverse multi-centered socially situated practices across time, space and institutional locations, then this has strong implications for classroom practice. Given any formulation of a mathematical theory, there are limitless ways of recontextualizing it as a school mathematics topic in the planned curriculum. Each "reading" of mathematics will be contingent on who does it, their purposes, the social context, the underlying ideological presuppositions prevalent, and who pays for the work. In addition, every realization of the taught mathematics curriculum is potentially unique for the presentation has elements of performance in it, and live interactions between teacher and students, that can never be repeated identically, even if the same textbook selections, chalkboard notes, worksheets, are used again. Moreover, the learned curriculum includes widely varying outcomes because of the idiosyncratic process of appropriation and personal construction of meaning by learners. In Derrida's understanding, the planned will not necessarily become the taught. Neither will the taught necessarily become the learned mathematics curriculum.

Mathematical knowledge is not analogous to a material entity that can be transmitted from one person to another, or delivered like a book or computer disc. It depends on a dual relationship between signifiers and a meaning-bearing context (conversation and its social context). Signs and rules can, in the extreme case, be "transmitted," that is, directly communicated, although this is not an effective teaching strategy (Seeger & Steinbring, 1990). The problems of teaching solely by exposition and direct instruction, and learning solely by rote and the practice and reinforcement of skills are well known (Ausubel, 1968; Bell, Costello, & Kuchemann, 1983; Skemp, 1976). The key issue is that meanings have to be constructed by each human being in turn, and the making of meanings by a person always draws upon their active (if unconscious) mobilization of existing elements of meaning and understanding, within a social context. These are enduring insights offered by cognitive, constructivist and social theories of learning, irrespective of any controversies between them (Burton, 1999; Ernest, 1994b; Steffe & Gale, 1995).

Great pedagogical knowledge and skill are needed for the successful teaching of a mathematical topic. It needs to be presented within a sup-

portive social context using a variety of representations and tasks with sufficient redundancy so that rules and relationships are inferred and
meanings constructed and elaborated. This normally takes place during an
asymmetric conversation between a teacher and a class of students in
which burgeoning capabilities are elicited and assessed.

Typically a teacher wishes the learner to learn some general item that is
applicable in multiple and novel situations, such as a mathematical concept, rule, generalized relation, skill or strategy. However if this item is presented explicitly as a general statement, often what is learned is precisely
this specific statement, such as a definition or descriptive sentence, rather
than a human capability. As such, it loses its generality and functional
power. In order to communicate the content more effectively it needs to
be embodied in specific and exemplified terms, routinely in a sequence of
relatively concrete examples so that the learner can construct and observe
the pattern and generalize it, as part of an underlying meaning structure.
Paradoxically general understanding is achieved only through experience
of concrete particulars. As in the "Topaze effect" (Brousseau, 1997), the
more explicitly the teacher states what the learner is supposed to learn, the
less possible the learning becomes.

A further complexity arises because the mathematical topics under consideration by learners change and develop over the course of their learning careers. A planned topic area may be intentionally and systematically
changed by the teacher, such as when a new topic area encompasses and
subsumes a previously learned topic. But sometimes the growth of school
topics involves the negation of existing rules and meanings, through the
adoption of new rules that contradict one or more of the old rules. For
example, to a young child the task $3 - 4$ is impossible. But later it has a
determinate answer: $3 - 4 = -1$. Similarly 3 divided by 4 $(3/4)$ is at first
impossible. Later it is not only possible, but $\frac{3}{4}$ names the answer to it, that
is, becomes a new kind of semiotic object, a fractional numeral.

These rule changes are necessitated by changes in the underlying operational meaning. Subtraction is often first understood in enactive/meta-
phoric terms as resulting from the partitioning of a collection of concrete
objects and the removal of one part. Hence $3 - 4$ *is* impossible. Subsequently subtraction is understood structurally as the inverse of addition on
an enlarged, more abstract domain of number. Hence since $3 - 3 = 0$, $3 - 4$
$= -1$. Very likely the later abstract meaning of subtraction cannot be developed without the earlier concrete meaning, so the apparent contradiction
is unavoidable. These changes and the problems they cause, have been
named epistemological obstacles (Brousseau, 1997; Sierpinska, 1994). The
process involves the student relinquishing meanings and rules already
learned before further progress might be made. There always needs to be

a process of negotiation between participants that may move toward a consensus of interpretation, never forgetting that there is also power at play.

The fragmented, multiple self also has implications for learning and the problem of transfer of learning from one context to another. Identity is a contingent historical construction dependent on personal and social developments, and within a person's school career multiple identities will be developed. The "self as learner of mathematics," despite the focused interest of mathematics educators, is by no means the dominant identity for a developing person. But when something goes awry in this formation (and this is neither common nor uncommon), negative attitudes and dispositions toward mathematics seem to result, which may or may not interact with other identities being formed. Other identities such as "self as boy/man," "self as girl/woman" are being formed and are important with strong presences outside as well as inside school. Much of the research on gender differences in mathematics can be understood in terms of the interaction of these developing identities (Walkerdine, 1998).

Within the mathematics classroom the development of identity as "self as learner of mathematics" is achieved through a historical process employing text-based tasks and mathematical conversations in which the symbolic comes to dominate. In learning to maintain the depersonalized, objectified and standardized discursive style of mathematics, the learner is also subjectifying herself, that is, constructing a limited and new self-identity as a mathematical subject (Ernest, 2003). This is a potentially rich area for exploration, one made possible by the postmodern conception of multiple selves. It is one of many dimensions of postmodernism and the subject of mathematics that is pregnant with further possibilities.

CONCLUDING THOUGHTS

In this chapter I have developed the idea of mathematics as conversation. I traced an anti-essentialist position by outlining how no predetermined feature dictates the emergence of mathematics, past, present or future. That perspective was set up against modernist meta-narratives that have tended to conceal the socially produced origins of mathematics. In arguing that language was fundamental to mathematics, I suggested that mathematics consists of language games with deeply entrenched rules and patterns that are very stable and enduring, but which always remain open to the possibility of change. As they change, so does the range of possibilities discovered within a mathematical system. In my mind, social constructivism expresses these ideas.

Mathematical knowledge, in social constructivist thinking, is not fixed, determined and ready-made. Rather, it comprises texts made and received by persons within social/institutional settings—persons with their own histories, expectations and interpretations. Mathematics is, at heart, conversational and hence indissolubly linked to its context. And because of that, it is always shaped and delimited by the contexts of making or utterance and those of reception or interpretation.

REFERENCES

Anderson, R.J., Hughes, J.A., & Sharrock, W.W. (1986). *Philosophy and the human sciences.* London: Croom Helm.

Ausubel, D.P. (1968). *Educational psychology, a cognitive view.* New York: Holt, Rinehart and Winston.

Bell, A.W., Costello, J., & Küchemann, D. (1983). *A survey of research in mathematical reducation. Part A: Teaching and learning.* Windsor: NFER- Nelson.

Bishop, A.J. (1988). *Mathematical enculturation.* Dordrecht: Kluwer.

Blake, W. (1796). Songs of experience (Reprinted with Songs of innocence). Oxford: Oxford University Press.

Bloor, D. (1984). A sociological theory of objectivity. In S.C. Brown (Ed.), *Objectivity and cultural divergence. Royal Institute of Philosophy lecture series* (Vol. 17, pp. 229–245). Cambridge: Cambridge University Press.

Bloor, D. (1991). *Knowledge and social imagery* (2nd ed.). Chicago: University of Chicago Press.

Brousseau, G. (1997). *Theory of didactical situations in mathematics.* Dordrecht: Kluwer.

Burton, L. (Ed.) (1999). *Learning mathematics: From hierarchies to networks.* London: Falmer Press.

Cooper, B. (1985). *Renegotiating secondary school mathematics.* Lewes: Falmer Press.

Davis, P.J., & Hersh, R. (1980). *The mathematical experience.* Boston: Birkhauser.

Derrida, J. (1976). *Of grammatology* (Trans: G. C. Spivak). Baltimore: The Johns Hopkins Press.

Descartes, R. (1637). *A discourse on method.* Translation in R. Descartes, *Philosophical Works*, Vol. 1. New York: Dover Press (1955).

Ernest, P. (1989). *The mathematics related belief systems of student primary school teachers.* Presentation at the 13th conference of the International Group for the Psychology of Mathematics Education, Paris, July 8–13.

Ernest, P. (1991). *The philosophy of mathematics education.* London: Falmer Press.

Ernest, P. (1994a). The dialogical nature of mathematics. In P. Ernest (Ed.), *Mathematics, education and philosophy: An international perspective* (pp. 33–48). London: Falmer Press.

Ernest, P. (Ed.) (1994b). *Constructing mathematical knowledge: Epistemology and mathematics education.* London: Falmer Press.

Ernest, P. (1998). *Social constructivism as a philosophy of mathematics.* Albany: SUNY Press.

Ernest, P. (2003). The epistemic subject in mathematical activity. In M. Anderson, A. Saenz-Ludlow, S. Zellweger, & V.V. Cifarelli (Eds.), *Educational perspectives on mathematics as semiosis: From thinking to interpreting to knowing* (pp. 81–106). New York: Legas Publishing.

Evans, J. (2000). *Adults, emotions and mathematics*. London: Falmer Press.

Foucault, M. (1972). *The archaeology of knowledge* (Trans: A. Sheridan). London: Tavistock.

Foucault, M. (1981). *The history of sexuality* (Part 1) (Trans: R. Hurley). Harmondsworth: Penguin Books.

Gödel, K. (1931). Über formal unentscheidbare sätze der principia mathematica. Translation in J. van Heijenoort (Ed.), *From Frege to Gödel: A source book in mathematical logic* (pp. 592–617). Cambridge, MA: Harvard University Press (1967).

Habermas, J. (1981). *The theory of communicative action* (2 Vols), Frankfurt am Main: Suhrkamp Verlag (Trans: T. McCarthy). Cambridge: Polity Press (1987 & 1991).

Harding, S. (1986). *The science question in feminism*. Milton Keynes: Open University Press.

Harré, R. (1983). *Personal being*. Oxford: Blackwell.

Heidegger, M. (1962). *Being and time* (Trans: J. Macquarrie & E. Robinson). New York: Harper and Row.

Henriques, J., Hollway, W., Urwin, C., Venn, C., & Walkerdine, V. (1984). *Changing the subject: Psychology, social regulation and subjectivity*. London: Methuen.

Hersh, R. (1997). *What is mathematics, really?* London: Jonathon Cape.

Hume, D. (1739). *A treatise of human nature, Book 1* (D. Macnabb, Ed., 1962, Glasgow: W. Collins, Fontana).

Kitcher, P. (1984). *The nature of mathematical knowledge*. Oxford: Oxford University Press.

Kline, M. (1980). *Mathematics the loss of certainty*. Oxford: Oxford University Press.

Lacan, J. (1977). *The four fundamental concepts of psycho-analysis*. London: Peregrine Books.

Lakatos, I. (1976). *Proofs and refutations*. Cambridge: Cambridge University Press.

Lakatos, I. (1978). *Philosophical papers* (2 Vols), Cambridge: Cambridge University Press.

Lave, J., & Wenger, E. (1991). *Situated learning: Legitimate peripheral participation*. Cambridge: Cambridge University Press.

Lemke, J. (1998). Multiplying meaning: Visual and verbal semiotics in scientific texts. In J.R. Martin & R. Vell (Eds.), *Reading science* (pp. 87–113). London: Routledge.

Lemke, J. (2003). Mathematics in the middle: Measure, picture, gesture, sign, and word. In M. Anderson, A. Saenz-Ludlow, S. Zellweger, & V.V. Cifarelli (Eds.), *Educational perspectives on mathematics as semiosis: From thinking to interpreting to knowing* (pp. 215–234). New York: Legas Publishing.

Lyotard, J.-F. (1984). *The postmodern condition: A report on knowledge*. Manchester: Manchester University Press (French original published by Les Editions de Minuit, 1979).

Mead, G.H. (1934). *Mind, self and society*. Chicago: University of Chicago Press.

Peirce, C.S. (1931–58). *Collected Papers* (8 Vols). Cambridge, MA: Harvard University Press.

Popper, K. (1970). *Objective knowledge*. London: Hutchinson.

Restivo, S. (1992). *Mathematics in society and history*. Dordrecht: Kluwer.

Restivo, S., Van Bendegem, J.P., & Fischer, R. (Eds.). (1993). *Math worlds: Philosophical and social studies of mathematics and mathematics education.* Albany: SUNY Press.

Rorty, R. (1979). *Philosophy and the mirror of nature.* Princeton, NJ: Princeton University Press.

Rotman, B. (1993). *Ad infinitum and the ghost in Turing's machine: Taking god out of mathematics and putting the body back in.* Stanford, CA: Stanford University Press.

Rotman, B. (1994). Mathematical writing, thinking and virtual reality. In P. Ernest (Ed.), *Mathematics, education and philosophy: An international perspective* (pp. 76–86). London: Falmer Press.

Rotman, B. (2000). *Mathematics as sign: Writing imagining, counting.* Stanford, CA: Stanford University Press.

Ryle, G. (1949). *The concept of mind.* London: Hutchinson.

Seeger, F., & Steinbring, H. (Eds.) (1990). *Overcoming the broadcast metaphor.* Bielefeld: Institute für Didaktik der Mathematik, Universität Bielefeld.

Shotter, J. (1993). *Conversational realities: Constructing life through language.* London: Sage.

Sierpinska, A. (1994). *On understanding in mathematics.* London: Falmer Press.

Skemp, R.R. (1976). Relational understanding and instrumental understanding. *Mathematics Teaching,* No. 77, 20–26.

Skovsmose, O. (1994). *Towards a philosophy of critical mathematics education.* Dordrecht: Kluwer.

Steffe, L.P., & Gale, J. (Eds.) (1995). *Constructivism in education.* Hillsdale, NJ: Lawrence Erlbaum Associates.

Tiles, M. (1991). *Mathematics and the image of reason.* London: Routledge.

Tymoczko, T. (Ed.) (1986). *New directions in the philosophy of mathematics.* Boston: Birkhauser.

Volosinov, V.N. (1973). *Marxism and the philosophy of language.* New York: Seminar Press.

Vygotsky, L. (1978). *Mind in society.* Cambridge, MA: Harvard University Press.

Walkerdine, V. (1998) *Counting girls out* (2nd ed.). London: Falmer Press.

Wenger, E. (1998). *Communities of practice: Learning, meaning and identity.* Cambridge: Cambridge University Press.

Wittgenstein, L. (1953). *Philosophical investigations* (Trans: G.E.M. Anscombe). Oxford: Basil Blackwell.

Wittgenstein, L. (1978). *Remarks on the foundations of mathematics* (rev. ed.). Cambridge, MA: Massachusetts Institute of Technology Press.

CHAPTER 3

POSTMODERNISM AS AN ATTITUDE OF CRITIQUE TO DOMINANT MATHEMATICS EDUCATION RESEARCH

Paola Valero

ABSTRACT

Postmodernism can be understood as a critical attitude towards the dominant discourses of mathematics education rooted in modern ideas. Using an example from my research on the processes of change in school mathematics education, I illustrate the discourse constructed in research around the mathematics learner and argue for the necessity of challenging that discourse and finding alternative ways of conceiving students. Those alternative conceptions open new possibilities of analysis in the direction of a postmodern critical mathematics education.

INTRODUCTION

When starting the writing of this chapter, I asked myself a simple question: What does it mean to adopt postmodern ideas to inform my practice as a

Mathematics Education Within the Postmodern, pages 35–54
Copyright © 2004 by Information Age Publishing
All rights of reproduction in any form reserved.

researcher in mathematics education? One possible answer is to provide evidence that my theoretical framework is based on work of postmodern authors. This is what some researchers in this volume have started to do in the last decade; their work has provided new insights for understanding essential aspects of mathematics education. However, this clear adoption of postmodern ideas is not what I have done in my research work. I see at least another possible answer to my initial question. I can try to make as explicit as possible how my activity as a researcher is permeated by a desire to question what has been taken for granted in mathematics education research. I am concerned with being postmodern by adopting an *attitude*. Such an attitude is a critical position expressed through my commitment to the examination of the dominant constructs and ways of generating knowledge that define the discipline in which my academic endeavor is inscribed.

My intention here is to show how my postmodern attitude emerged and consolidated as I faced the multiple challenges of researching reform in school mathematics education from a sociopolitical perspective (see Valero, 2002). In other words, my postmodern attitude did not result from a conscious paradigm selection; rather, it was constructed as I met school leaders, teachers and students in different schools in the world, whose lives shook me in significant ways. In particular, in becoming postmodern, I found ways to challenge the dominant discourse and constructs of mathematics education research at the base of explanations about change in mathematics education practices. The examination of this discourse through examples of my research will lead me to formulating what I see as the task of a *postmodern critical mathematics education*.

Let us start with my meeting some students in Colombia.

BUT WHY SHOULD I STUDY IF...

When Julia asked me to take care of her 10th grade class for one session, I felt excited. I wanted to explore students' perceptions about their learning, their class, and their mathematical experience. We agreed on dividing the session in two parts: One for talking about their perceptions of mathematics, and the other for finishing the worksheet that they had started in the previous lesson.

> **Paola**: Guys, why don't we make a round table? That would be the best organization for what we are going to do today.
>
> **Students**: Where is Julia?
>
> **Paola**: She could not come today because she is handing in an application for a project in mathematics. Your teacher

works hard! OK. As Julia told you, I am in this school observing the way your teachers teach mathematics, and understanding how they work and why they do what they do. I would like to have a chat with you about how you see this story of learning mathematics. My intention is that both you and I ask questions of each other and that we have a nice conversation. So, to start, do you like mathematics?

[Silence]

    Paola: Opinions? Is it nice? Is it boring?

[Silence]

    Paola: Nobody dares to say anything?

[Silence]

    Paola: I won't tell Julia about what you say here. This is between you and me!

  Carolina: Well, I think most of us think it is difficult, and therefore we don't like it.

    Paola: Why do you think it is difficult?

[Silence]

    Tomás: It's just that it can't be used for anything except for passing the test!

    Paola: If it is difficult and you can't see its use, does it mean that it is not worth studying?

[Silence]

    Daniel: Well, if we don't study, we don't pass. And if we don't pass, we don't get our high school diploma, and that is the minimum thing one needs to get a job, even in the supermarket across the street.

    Paola: So anyway, it is important to learn mathematics!

[Silence]

    Tomás: It is not the mathematics! It is passing, getting out of this school and getting a job!

[Silence]

The conversation went on like this for a while until it was time to continue with the worksheet. Students' almost reticent involvement in our conversation differed from their apparent involvement in solving the trigonometry exercises in the worksheet. I especially noticed two boys, who looked at the paper and laughed, looked at me and laughed, looked at the others and laughed. They called me for help:

    Andrés: Do you know the answers? Can you do this?

    Paola: Well, I don't know it by heart, but we can try to do it together.

> **José**: Do you really like this? How come? I can't understand you! But, tell us, why are you really here in this school?
>
> **Paola**: As I told you, I want to understand the way teachers teach mathematics and, of course, I had to come here and talk to the teachers, see what they do, and talk to students like you, as well.
>
> **Andrés**: Ah! But why did you decide to come to this school if there are so many good ones ... some that might be closer to you.
>
> **Paola**: I think that good schools are not interesting. I wanted to see a school that is normal, you know, not good nor bad, and I think that this one is precisely what I wanted to see!

They were curious to know my intentions and motivations for being in their school. They wanted to know about my life, where I have studied, and why I was living outside Colombia. They couldn't understand why I was there in that "poor" school, talking to them if:

> **Andrés**: We can see in your face that you have never suffered; you've got it easy.

I have not suffered, that is true. But that did not mean that I had got it easy. I studied hard to have the chance of doing what I was doing. I intended to tell them that there were reasons to study and to be interested in school and maybe in mathematics. But they could not see it:

> **José**: The only class I would like to pay attention to is English because I want to get out of this fucking place and go to the USA. Though, I don't even manage to say "Hello, good morning."

MEETING STUDENTS IN THE MATHEMATICS CLASSROOM

This episode took place during my visit to Esperanza[1] Secondary School, a public school located in a working class area of Bogotá, Colombia. The episode captured my attention. A first striking circumstance was the students' silence to my invitation to talk about their experience with school mathematics. Although the students were familiar to me, there could have been many reasons for their blockage: fear that I might tell their teacher what they had said; reaction toward the power of a "foreign" adult researcher; passivity and lack of self confidence to express their opinions; lack of clarity in my questions ... I did not explore in detail why the interaction with the whole group went that way because the chat with Andrés and José captivated me.

These two boys seemed to be interested in inquiring of me, but not to be inquired by me. They clearly drew a line between them as members of the school and of a community and me as an outsider. I interpreted their *words. We can see in your face that you have never suffered; you've got it easy* as a marking of me as belonging to a "world" that differed from that of people who have to struggle. It was the only time during my stay in Esperanza School when somebody challenged me in this direct way. I questioned whether my intention of making students feel interested in their studies, especially in mathematics, was meaningful to them. Probably they could not see any reason to do so....

I turned to research literature in search for plausible interpretations about the relevance of this episode as an exemplary situation of practice that may occur in a mathematics classroom where change is "in the air." I made a systematic scrutiny of literature published in journals and books in the English language since the beginning of the 1990s. My search soon became the identification of those ideas that seemed to be central to defining students of the reform. The notions that researchers have used to express the qualities of these people seemed to constitute a portrait of them. That is how students are portrayed in mathematics education research.

TRACING DOMINANT REFORM DISCOURSE ON THE MATHEMATICS LEARNER

One of the main characteristics of the current reform discourse, as put forward by the research literature, is the placement of students and their mathematical thinking development as the target of the whole educational enterprise. This discourse portrays students mainly as *cognitive subjects* around which all learning and for whom all teaching happens. There are different reasons for this. Recent reform ideas have, in the main, adopted Piaget's genetic epistemology and more recently Vygotsky's sociocultural epistemology to give an account of human learning. Numerous researchers have recontextualized these theories into mathematics education (e.g., Lerman, 1996; Nunes, 1992; von Glasersfeld, 1995). Researchers involved in rethinking mathematics learning from (socio)constructivist and sociocultural perspectives have placed an emphasis on the student as a significant unit around which learning happens. The way in which students come to develop mathematical thinking and meaning is the central aim of the whole reform discourse.

There have also emerged associated research trends that concentrate on different characteristics of the mathematical cognizing subject. One tendency has explored conceptual development (e.g., Sfard, 1991). This type of research has illuminated how students construct mathematical knowledge and develop higher order mathematical thinking. Another trend has emphasized the different sources of meaning making in learning (e.g., Cobb, 2000). Other trends have explored diverse factors connected with mathematical learning. Many studies consider the influence of the affective domain on mathematical learning as seen in research dealing with motivation for, attitudes toward, and beliefs about mathematics (e.g., McLeod, 1992). Research has also paid attention to the interaction aspects that affect mathematical thinking development, such as peer-student cooperation (e.g., Cobb, Boufi, McClain, & Whitenack, 1997), the classroom culture (e.g., Voigt, 1996), and parents' involvement in students' learning (e.g., Desimone, 1999). Furthermore, individual characteristics connected with broader sociocultural structures have also been considered as factors influencing the student's possibilities to participate in mathematical learning. Studies focusing on gender (e.g., Keitel, 1998) and social class (e.g., Zevenbergen, 2000) have complemented the view of mathematics education in the process of change in mathematics education.

From these trends that describe different aspects of students we can conclude that the dominant reform discourse portrays students as *cognitive subjects*, representatives of the human race, whose cognitive development can be described in terms of standard mental processes. Of course there has been a growing body of literature emphasizing cultural, sociological and political perspectives (Lerman, 2000); nevertheless this is minor in comparison to what constitutes most research. Most mathematics education research talks about the universal, normal child and how she thinks mathematically. But, what is the problem? you may ask. Mathematics education is about mathematical learning and there is no learning without an emphasis on either the learner or the mathematics. *My* argument goes in the direction that there is much more at stake than that kind of learner when talking about mathematics education.

CHALLENGING THE DISCOURSE

If I were asked to draw the "reform student" I would paint a being that looks like an outer-space visitor, with a big head, probably a little heart, and a tiny chunk of body. That being would be mainly alone and mostly talk about mathematics learning, and would see the world through his school mathematical experience. That would be a "schizo[2]-being" since she has a clearly divided self—one that has to do with mathematics and the other that has to do with other unrelated things.

This caricature of the "schizomathematicslearner" was seriously questioned when Andrés and José challenged me. Is it true that Andrés and José are "schizomathematicslearners?" Is it true that I can conceive of them as students who are mainly learning mathematics? Is it true that they act in one particular form? Is it true that they are interested in learning mathematics? Is it true that they can separate their experience in learning mathematics from their whole school experience and from their whole life experience? Their words, *The only class I would like to pay attention to is English because I want to get out of this fucking place and go to the USA. Though, I don't even manage to say "Hello, good morning"* made me think that students are not only mathematics learners, are not necessarily interested in learning mathematics, do not act in the ways expected, do not perceive a separation between their mathematical learning experience and their whole school and life experience, and have a whole human integrity. I was forced to broaden the focus of what seems to be fundamental in mathematics education research concerning students in schools. In search of ways to broaden my view about students in mathematics classrooms, I examined the research literature again in light of José and Andrés and their context.

Many research papers depict classroom situations in which students perform mathematical activity. These classrooms are neat; students behave well—for one thing, they never swear—and teachers never deal with discipline problems, lack of attention or insufficient motivation. As an example of such a portrait, let us examine the description of a classroom sequence, in the developmental research of Cobb and his colleagues; a 7th grade class was working with statistical data analysis (Cobb, 2000). Researchers collaborating with the teacher in the class were concerned to make students' conceptions of data analysis evolve in a desired direction that is considered more mathematically rich in understanding. The episode happened in an environment in which the thirty students in the class had a computerized, mini-tool for data handling in order to manipulate the information provided by two treatment protocols for AIDS. Students engaged in a series of activities such as accepting the task, comparing the information provided, partitioning data, observing it, analyzing it, concluding with respect to the task based on the information, writing reports, discussing their results, and arguing for their suitability. From the description it seems that all the students were engaged in the task and participated in the situation. There were no students who failed to achieve the expected aim. No one had problems with the fact that they were deciding about a treatment for AIDS. All followed the game proposed by the teacher of deciding which data showed the more efficient treatment. Had the students been South African, would engagement in the task have been so "unproblematic?"

This description is an unproblematic representation of the students, of their engagement and interest in mathematics learning. Cobb's "clean" classroom does not fit with the diversity that one finds in real classrooms.[3] Certainly it did not fit with the environment in which Andrés and José lived, where among forty students, there were some who participated, some who remained quiet, and some who were talking about other topics. I observed the class for some weeks when Julia, the teacher, developed the topic of trigonometrical identities. In order to provide the students with a richer understanding of the topic, she implemented a sequence where graphic and symbolic representations of the identities were combined to show the equivalence of, for example, $cosec^2\ \alpha$ and $cot^2\ \alpha + 1$, in the case of the identity $cosec^2\ \alpha = cot^2\ \alpha + 1$. In that sequence only a few students contributed to the discussion. Andrés and José were among the silent majority who hardly ever participated. Sitting among the students while the teaching was happening, I noticed that students did other things. Chatting about the football match the night before or commenting on the last episode of the most popular soap opera were parallel activities in the classroom (Valero, 2002). This diversity in students' participation is not unique. What might then lie within the scope of students' interests when being within a mathematics classroom?

Mellin-Olsen (1987) focused his social pedagogy of mathematics on students' possibilities for action in relation to school mathematics. He explored what might be at the base of students' activity, from the point of view of the students. At a time when the core ideas of the constructivist reform were in formation, Mellin-Olsen highlighted an important issue:

> [A]s educators we see them [students] decide to learn or not to learn. As educators we are cheating ourselves if we do not make this phenomenon central to our theorizations. (p. 157)

Students may decide not to learn. Mellin-Olsen directs our attention to the fact that students' activity should not only be described in cognitive terms, but also and mainly in sociological terms. The activity of a person—even if it is mathematical—is a result of the person's interaction with the immediate environment. That environment is a field where the person meets other people and social institutions. The individual expresses personal ideological constructions about how to act in the environment, and is exposed to the expression of others' ideologies and of broad social ideologies. From these interactions emerge rationales for behavior, which belong to the individual but are constructed in the field of social interaction. These rationales provide justifications for individual activity.

This point of view also allows us to tackle the portrait of "schizomathe-maticslearners" as context-free human beings. I contend that a great deal of research produced during the 1990s, which fed the constructivist reform trend, adopted views of context restricted to *task context* and to *inter-action context* (see Valero, 2002; Vithal & Valero, 2003). Let me examine the implications of these views of context on the issue of the representation of students as context-free beings.

In constructivist research, where the notion of context is mainly associ-ated with the task-context, students are not taken to be concrete human beings in a particular social and historical situation, but beings who repre-sent an abstract "human being." Through the study of one, some, or many children, we can understand the thinking processes of the universal child. The emphasis on describing the mental processes that these "abstract stu-dents" follow has the effect of detaching children from the frames in which concrete human beings exist. As an example, let us examine the text writ-ten by Maher and Martino (1996). The researchers followed a group of children for three years in order to study "how children build up their ideas when confronted with problematic tasks that promote thoughtful-ness about mathematical situations" (p. 342). They concentrated on the performance of two students about whom the only information available is their names, gender, and that they were in 1st and 3rd grades at the time at which they were observed. From their analysis, Maher and Martino draw general conclusions about children's mathematical learning. What is important for the authors about these children is how they come to think about the piece of mathematics displayed by the problem that they face, and how the individual and inter-individual interactions with problems advance children's knowledge construction. The paper, like many others, does not mention the place where the experiment took place. Although the mathematical problem itself is given a context, these "laboratory chil-dren" are modeled as context-free universal children.

Social-constructivist and interactionist research reveals more of the con-text of the children. The context in this case is the classroom and children get reified as "classroom bounded beings." We can look again at the paper by Cobb (2000). The integration of a psychological with a social perspec-tive allows Cobb to embrace all aspects of individual learning as well as of the inter-subjective, thereby including social interaction in the frame of the classroom. This combination led him to the formulation of a model in which students' mathematical learning is situated with respect to the class-room micro-culture. In this micro-culture, individual interpretations and reasoning emerge simultaneously with the development of the classroom's social norms, its sociomathematical norms, and its mathematical practices. He concludes that the mathematical practice that emerged in the class-room was constituted through the public discourse made explicit in the interaction between students and teacher. The paper does not provide information about whom the students are, except for their age, grade and names. Although the reader is provided with information about them as participants of a classroom situation where individual and social mathematical activities are taking place, there is no reference to participants as embodied learners—learners living and hence, knowing, within the larger social context.

Sociocultural studies of learning from anthropological perspectives put forward the idea of context in mathematics education as the *situation context*. Many of the researchers who claim to have followed this trend have redefined the possible meaning of students' context in terms of the *settings* for mathematical activity (e.g., Boaler, 1997). As an example of this trend let us examine Boaler and Greeno (2000). They show how students in two different settings (didactic teaching and discussion-based teaching) participate in social practices which lead them to hold almost opposite ideas of what a mathematics classroom is, of their identities as mathematics learners and as potential "mathematicians', and of their capacity as individuals to contribute to the social world of the classroom. They conclude that this approach allows them to see mathematics learning as a process in which individuals "learn to *be*" (p. 172). Students in this kind of discourse are presented as human beings within the context of a community. Boaler and Greeno provide some information about their informants. They are high-achieving students, placed at the top of the mathematics stream. Their middle and upper-middle socioeconomic background is clear, as well as the whole socioeconomic status of the schools. Many of these students have possibilities to continue with college education. All of these characteristics, which may be related to the students' instrumental intentions, are not analyzed as factors influencing the students' construction of their identities. Somehow it seems that once the broader context of the students is stated, it is left aside and it slowly dilutes in the characteristics of the closed social context defined by the mathematical learning community of practice. Despite the enlargement of the notion of the students' context here, it seems that the elements of the *sociopolitical context* of the students, as fully historical members of a large society, dilute when participating in a mathematics community of practice. The different discursive constructions mentioned above oppose the nature of the students that one meets in real classrooms. Andrés and José are two full, social beings in a particular historical time, geographical location, and social position. The characteristics of their sociopolitical context cannot be discarded when considering their mathematical learning experience. Understanding the situatedness of students' intentions about participation in mathematical learning implies keeping a clear connection between the students' micro-context (the task, the interaction, and the situation context) and their broad macro-context as members of a particular society at a given time.

Let us look closely at Andrés and José. They confronted me by marking a difference between my socioeconomic position and theirs. One possible interpretation of this fact is that they perceive that we have different histories, motivations and attitudes toward schooling and school mathematics, given who we are. For a person like me, from a middle class background, it could make sense to have a strong instrumental rationale for getting involved in learning school mathematics. I could gain upward social mobility by being educated. At the time I was in 10th grade, getting good (math) qualifications meant opening doors to the future. Getting a high score in the National High-stakes Examination secured entry into a prestigious university. Actual enrollment depended on your parents being able to afford the high tuition fees. That was my case. I talked with encouraging words to Andrés and José from my experience. I wondered how they linked my encouragement to their own futures.

Information about the school population in Esperanza's Institutional Educational Project[1] (IEP) illustrates the background of students attending the school. Most students are raised within single-mother, divorced or second-union families. Most parents have finished primary education (5th grade); some have reached a secondary level (11th grade), and only a few have university or college degrees. Most families live on an income equivalent to the legal minimal salary, estimated, at the time of the research, to be around $US100 monthly. A third of the school population lacks a basic level of nutrition for the required mental and physical development at their age. Only 10% of the whole student population is reported to be completely healthy. Moreover, a fifth of the student body needs to work during the shift when they don't attend school or during the weekends as a way of helping support the family (Esperanza, 1997, section 1.5). Esperanza's IEP concludes that students' families are placed in a low and low-middle socioeconomic level.

I cannot provide concrete evidence about how Andrés and José's positioning affects their perception of schooling or school mathematics. However, I assume that such a relation exists. Their whole participation in school mathematics is strongly influenced by whom they are, to which cultural group they belong, and by how the future looks from that position in society. This allows me to think about the lack of separation of students' learning in a micro-context such as the classroom from the broader context in which students' lives develop.

Esperanza's IEP also reports the results of a questionnaire inquiry carried out among the whole population of parents and students in the school in 1996. The results provide an insight into what parent and students see as possibilities in their community. The questionnaires asked both parents and students about the technological education[5] that the school had planned to implement as part of its long-term mission of contributing to the improvement of the educational community's life quality. Parents were asked about: the kinds of technological instruction that could be relevant for their children, the types of jobs available in the community, and the kinds of employment that they could offer students finishing high school in Esperanza. Parents said that the most relevant technical education to be offered is trade and informatics. They saw that the locality could offer jobs for traders, salesmen, employees in banks, and, in a minor proportion, preparation for university studies. They could offer jobs to students mainly as salesmen, secretaries, accountants, and cashiers.

Students were asked about the kinds of technological formation that fulfill their interests and needs, the possible uses of such an education, the job opportunities in the local community and their expectations after finishing high school. The students favored informatics and trade. The locality could offer them jobs as salesmen, traders, cashiers, bank employees, and secretaries, yet few reported expectations to work as traders, with computers, or as secretaries.

Students saw that they could use their current education to enter university, whereas parents did not consider higher education as a possibility. Julia and Laura, the teachers in charge of mathematics in grades 10th and 11th, discussed with me how, despite the fact that some students wished to follow university studies, few of them could actually make it. For that particular group of students leaving 11th grade in 1999, the scores in the National High-stakes Examination (one of the passports into university) were not very high. Another factor was the high cost of higher education tuition fees which, the teachers emphasized, at that particular moment few families could afford to pay. Even worse, students could not enter university since the family survival depended on them getting a job to contribute economically to the family. The situation, as Julia and Laura commented, was very discouraging.

We also need to consider the predicaments facing Colombia at the end of the 1990s. As the result of a confluence of economic and politic factors, Colombia entered a crisis that has worsened peoples' life possibilities. One of the immediate causes of the economic recession was the combined national and international control of the flow of money stemming from drug traffic activities. There was, too, a scandal involving the President at that time. The combination of imminent internal political instability with the start of economic decay, generated panic in national and international investors who, little by little, withdrew their capital from the country. A decrease in the Balance of Payments, a sharp rise in the unemployment rate between 1998 and 2000, as well as the presence of guerrilla, extreme-right and paramilitary group activity, all contributed to the seriousness of the economic situation. At the turn of the century the possibilities of regaining both economic and political stability seemed remote. For many Colombians the only alternative seemed to migrate out of the country in search of future opportunities.

It is within this context that José's emotive comments emerged. It is fair to interpret those comments as an indication of the connection that exists for José between school experience and the larger context. Does it make sense to study in a place where there seems to be no future? That is probably the question revolving around in the heads of some of Esperanza School's students–or of some South African students in a Black township, or "second generation immigrant" students in a Turkish ghetto in Denmark, or a Palestinian student in the middle of the crossfire ... Mellin-Olsen's rationale of learning emerged in a context in which gaining qualifications represented a possibility for the future. But what if the future becomes unclear from the students' point of view?

Skovsmose (1994) introduced the concept of *intentionality* and formulated the issue of rationales for action in terms of the learner's intentions of learning. Without a conscious will for engaging in an action, there is no learning. A person's intentions for acting are connected with the person's dispositions. Skovsmose refers to dispositions in terms of the person's background and foreground. The background, he says, "can be interpreted as that socially constructed network of relationships and meanings which belong to the history of the person" (p. 179). The foreground refers to "the possibilities which the social situation makes available for the individual to perceive as his or her possibilities" (p. 179). Students' background and foreground are not connected linearly by a cause–effect relationship. Rather, the kind of relationship between background and foreground as sources of learning intentions depends on how the individual interprets both his personal history and his potential future in relation to a particular social situation.

In the case of the Colombian students, this means that students constantly weigh the choice of participating in schooling and school mathematics learning, taking into consideration their personal history and the possible future in a whole social context. Sometimes the background–foreground context is such that it may make sense for some to engage in mathematics education as a means of acquiring socially valued qualifications that are needed to secure a better future position. But on other occasions the combination shows a less optimistic landscape in which no engagement will improve the chances for the future. If a researcher is interested in exploring engagement, then the strong influence of the broad historic, social, political and economic context in which students live and generate intentions of action for learning cannot be ignored in mathematics education.

"REALIZING" STUDENTS

I questioned the dominant portrait of students as cognitive subjects, involved in the learning of mathematics and contextualized in the classroom. I proposed a "realized" view of students as whole learners, who have multiple motives for learning, and who live in a broad context which influences their intentions to participate in school mathematics practices.

Realized students are an essential characteristic of a sociopolitical perspective in mathematics education. There are several reasons for this. First of all, having in mind students who are full human beings allows us to recognize the *agency* that students have in the whole educational enterprise. Saying that students are agents means that they act in complex social situations. Students build different reasons to get involved in school mathematics practices. Their participation in those practices, which are perceived by the student as a whole social experience and not only as an intellectual, mental or cognitive endeavor, stems from their intention to act and influence the social space where the learning and teaching of mathematics take place. In other words, students are participants in a social situation, and the development of that social situation depends strongly on the agency that they can exercise in it.

Second, students gain a position of power in relation to the teacher and how school mathematics practices develop. In a real classroom situation, there are students who seem to follow the game set in place by the teacher. Those are the students whose intentions seem to "be aligned" with some of the teacher's intentions in a situation. These may be the well-behaved, bright students that many research reports show. But, in real classrooms, this type of student tends to represent only a small proportion of the class. There are others whose intentions diverge from the teacher's. They decide to participate in different ways: keeping silent, bullying, or resisting (e.g., Alrø & Skovsmose, 2002). These are groups of students who openly adopt an attitude of no-participation in the game proposed by the teacher. From a traditional mathematics education viewpoint, these would be considered "deviant" or "problematic" students who need to be normalized. From a sociopolitical outlook, these students are expressing their agency to engage (or otherwise) in learning and, therefore, research and practice from that perspective gives serious consideration to the intentionality behind their participation. This is the basis for a negotiation between teacher and students about how to evolve school mathematics practices. Even in the case of the "aligned" students the same exploration and negotiation are needed. The imposition of the teacher's agenda on students' intentions risks generating the situation of not meeting students in a place that they perceive as relevant to their school mathematics experience.

Mathematics education without open *negotiation of intentionality* risks being a failure.

Finally, if students are agents and negotiation can help bring their intentions into the educational scene, real *empowerment* may take place. Empowerment in the reform discourse has been presented as the capacity available to students through acquisition of the intrinsic power of mathematics. In most cases, teachers who "possess" it wish to transfer it to students. It is the knowing of something (mathematics) that confers power to students (and to teachers). This view of power internally founded in mathematics is problematic (Skovsmose & Valero, 2001, 2002). In the situation in which students are recognized as agents of the educational process, empowerment does not emerge from the "possession" of mathematics, but from the position that students adopt to influence the social practices where mathematics is taught and learned. Empowerment, then, is not passed from the teacher to the student by means of the transference of a "powerful knowledge." As sociocultural studies on learning have shown, knowledge may not be transferable (e.g., Lave, 1988). Sociologists and micro-political analysts have also argued that power is not an intrinsic characteristic of a person, but the manifestation of a relation in which people position themselves in order to influence the outcomes of a situation using diverse tools (Foucault, 1972). Empowerment needs to be defined in terms of the potentialities for students to participate in school mathematics practices. They get empowered when, through that participation, they position themselves in ways that are significant for the development of the practice. In that positioning their intentions, their negotiation with the teacher about them, and their actual involvement in actions is connected to mathematics. The learning about positioning themselves in diverse practices and using diverse resources that schooling offers to them is what constitutes empowerment in a school situation.

POSTMODERNISM AS AN ATTITUDE IN MATHEMATICS EDUCATION RESEARCH

I started this paper arguing that postmodernism can be understood as an attitude of critique toward existing dominant constructs of mathematics education research. I also indicated my intention to show what could be the meaning of a critical postmodern mathematics education. In order to exemplify how I have constructed such a postmodern attitude, I illustrated the questioning that emerged from my interaction with two students in a Colombian high school. That questioning led me to search for possible interpretations about the mathematics learner. In that search I realized that there were "ways of talking" about the student that appeared again and again in many research papers. I engaged in the task of evidencing the characteristics of the discourse that dominant mathematics education research has built around the student. I argued that the discourse of researchers has constructed a discursive object that is a distorted representation of the real human beings that we meet in schools. I also argued for the deconstruction of such an object and signaled some elements of an alternative way of conceiving students in our research endeavor.

At this point you may be wondering whether my critique attitude is, in reality, postmodern. Some authors have referred to postmodernism as a trend of thought and social theorizing that points to the crack of the modern project and of its foundations in an emerging social organization that challenges both the ideas of modernism and the existence of modernity itself. Kumar (1995, p. 67) phrases this idea in the following terms: "The end of modernity is in this view the occasion for reflecting on the experience of modernity; post-modernity is that condition of reflectiveness." As part of such a condition of questioning, some postmodern authors during the last 30 years have examined the ways in which the role of intellectuals has produced the discourse of science, and how that discourse in itself does not "talk about the world"—as if discourse, scientific knowledge and reality were located in separate realms of existence—but actually "constructs" the world. One preoccupation in postmodernism, then, is the way in which the scientific endeavor has contributed to the creation, consolidation and maintenance of modernism, and how a meta-reflection on science creation itself can allow us to understand the mechanisms that associate research and the construction of the world (see Foucault's, 1970, poststructuralist analysis of science). For mathematics education research in general, and for my definition of postmodernism as an attitude of critique, in particular, these ideas imply taking a look at the way in which mathematics education as a scientific discipline has phrased and constructed the "objects-subjects" of its study. This represents a change in focus from the foundations upon which dominant, empirical and experimental research about teaching and learning practices has been conducted in the field, to the inclusion of the way in which the discourse of the scientific community is implicated in the production of those practices. In other words, the practices of "practitioners" intermesh with the practices of "researchers," and the role of the researcher evidences their mutual constitutive character.

In my path to construct an attitude of postmodern critique, I have found clues in the work of Popkewitz and his colleagues (e.g., Popkewitz, 2002, forthcoming; Popkewitz & Brennan, 1998). Using the tools of Foucauldian theory in education, Popkewitz has put forward an analysis of mathematics education reform in the United States and the discourse that curriculum studies in mathematics has constructed. He sees mathematics education as a set of cultural practices that represent systems of reason through which governance is effected. In that sense, mathematics education is:

> … an ordering practice analogous to creating a uniform system of taxes, the development of uniform measurements, and urban planning. It is an inscription device that makes the child legible and administrable. The mathematics curriculum embodies rules and standards of reason that order how judgments are made, conclusions drawn, rectification proposed, and the fields of existence made manageable and predictable. (Popkewitz, 2002, p. 36)

In explaining the mechanisms for governing children, and hence teachers, Popkewitz points to the role of psychology as the translation tool that allows the alchemy of school subjects. Mathematics education research builds constructs that allow us to talk about the desired characteristics of students' behavior, toward which mathematics education practices should converge. This construction, through which the processes of normalization and exclusion at the core of modernity operate, is supported by the illusion that mathematics education is a matter of mathematical learning.

Popkewitz's analysis evidences the subtle mechanisms of power embedded in our task as mathematics educators, and invites a reflection on the assumptions behind our scientific endeavor. The constructs and approaches to practice dominant in mathematics education research, traditionally built on the intersection between mathematics and psychology (Kilpatrick, 1992), have made possible the creation of the ways of acting and ways of talking that guide both our activity as researchers and as mathematics teachers. The whole of our discourse, although very well intentioned, however, is constantly effecting power. The mechanisms of power that constitute the most intimate fibers of our still very modern society are being constantly reproduced and remain unquestioned when we adhere to the view of students as "schizomathematicslearners," all for the sake of better mathematical understanding.

We need to undertake a critique of the constructs and guides for practice that mathematics education research has produced for two reasons. The first is to support a search for plausible, alternative understandings of the social practices of mathematics education in schools; the second is to break with the deeply entrenched modern systems of reason in which our discipline has built. The road of critique in mathematics education has been paved with the recontextualization of some of the essential ideas of critical education into our field. Furthermore, we have also taken a step in the direction of effecting a critique of mathematics itself and its role in current societies and, therefore, of mathematics education in highly technological societies (e.g., Skovsmose, 1994). We are starting to take critique a step further by examining mathematics education research itself and by investigating how researchers engage in the production and reproduction of the mechanisms of power that have been central to the constitution of modernity.

ACKNOWLEDGMENT

I would like to thank Ole Skovsmose for his comments on previous versions of this chapter.

NOTES

1. The word "esperanza" in Spanish means "hope." This fictitious name for this school illustrates teachers' commitment to their work, and their hopes of contributing to the improvement of students' life conditions.

2. I am aware of the possible problems of using the prefix "schizo-" to construct this image. I use the prefix as "a combining form meaning "split," used in the formation of compound words: schizogenetic. [<Gr, comb. form repr. schízein to part, split]" (Webster's encyclopedic unabridged dictionary, 1996, p. 1714).

3. Vithal (2000) presents crucial descriptions of South African mathematics classrooms in which all the reality and mess of teaching are visible to the reader. So do Alrø and Skovsmose (2002) in their descriptions of some Danish classrooms.

4. The school's IEP is an extensive document produced as a result of the negotiation with Esperanza's educational community about the philosophical rationale for all educational activities in the school. After the Colombian General Law of Education in 1994, all Colombian schools were required to write their own IEP. Esperanza's IEP was finished in 1997 and it is entitled "The improvement of the educational milieu and its influence on students' quality of life" (Esperanza School, 1997).

5. Esperanza Secondary School offers an academic secondary and high-school curriculum, that is, the course of studies oriented to university access. Nevertheless, the school as part of its plans wanted to introduce a minor group of subjects that could also offer the student a vocational, practical and job-oriented education. That is, the school planned what is referred to as "technological subjects" (trade, drawing, informatics, recreation, and health).

REFERENCES

Alrø, H., & Skovsmose, O. (2002). *Dialogue and learning in mathematics education.* Dordrecht: Kluwer.

Boaler, J. (1997). *Experiencing school mathematics.* Buckingham: Open University Press.

Boaler, J., & Greeno, J. (2000). Identity, agency, and knowing in mathematics worlds. In J. Boaler (Ed.), *Multiple perspectives on mathematics teaching and learning* (pp. 171–200). Westport: Ablex Publishing.

Cobb, P. (2000). The importance of a situated view of learning to the design of research and instruction. In J. Boaler (Ed.), *Multiple perspectives on mathematics teaching and learning* (pp. 45–82). Westport, CT: Ablex Publishing.

Cobb, P., Boufi, A., McClain, K., & Whitenack, J. (1997). Reflective discourse and collective reflection. *Journal for Research in Mathematics Education, 8*(3), 258–277.

Desimone, L. (1999). Linking parent involvement with student achievement: Do race and income matter? *Journal of Educational Research, 93*(1), 11–30.

Esperanza School (1997). *The improvement of the educational milieu and its influence on students' quality of life.* Bogotá: Author.

Foucault, M. (1970). *The order of things: An archaeology of the human sciences* (Trans: A. Sheridan). London: Tavistock.

Foucault, M. (1972). *The archaeology of knowledge* (Trans: A. Sheridan). New York: Pantheon.

Keitel, C. (Ed.) (1998). *Social justice and mathematics education: Gender, class, ethnicity and the politics of schooling.* Berlin: IOWME—Freie Universität Berlin.

Kilpatrick, J. (1992). A history of research in mathematics education. In D. Grouws (Ed.), *Handbook of research on mathematics teaching and learning* (pp. 3–38). New York: Macmillan.

Kumar, K. (1995). *From post-industrial to post-modern society.* Oxford: Blackwell.

Lave, J. (1988). *Cognition in practice: Mind, mathematics and culture in everyday life.* Cambridge: Cambridge University Press.

Lerman, S. (1996). Intersubjectivity in mathematics learning: A challenge to the radical constructivist paradigm. *Journal for Research in Mathematics Education, 26*(2), 133–150.

Lerman, S. (2000). The social turn in mathematics education research. In J. Boaler (Ed.), *Multiple perspectives on mathematics teaching and learning* (pp. 19–44). Westport, CT: Ablex.

Maher, C.A., & Martino, A. (1996). Young children invent methods of proof: The gang of four. In L. Steffe, P. Nesher, P. Cobb, G. Goldin, & B. Greer (Eds.), *Theories of mathematical learning* (pp. 431–447). Mahwah, NJ: Lawrence Erlbaum.

McLeod, D.B. (1992). Research on affect in mathematics education: A reconceptualization. In D.A. Grouws (Ed.), *Handbook of research on mathematics teaching and learning* (pp. 575–596). New York: Macmillan.

Mellin-Olsen, S. (1987). *The politics of mathematics education.* Dordrecht: Kluwer.

Nunes, T. (1992). Cognitive invariants and cultural variations in mathematical concepts. *International Journal of Behavioral Development, 15*(4), 433–453.

Popkewitz, T. (2002). Whose heaven and whose redemption? The alchemy of the mathematics curriculum to save (please check one or all of the following: (a) the economy, (b) democracy, (c) the nation, (d) human rights, (d) the welfare state, (e) the individual). In P. Valero & O. Skovsmose (Eds.), *Proceedings of the Third International MES Conference* (pp. 29–45). Copenhagen: Centre for Research in Learning Mathematics.

Popkewitz, T. (forthcoming). School subjects, the politics of knowledge, and the projects of intellectuals in change. In P. Valero & R. Zevenbergen (Eds.),

Researching the socio-political dimensions of mathematics education: Issues of power in theory and methodology. Dordrecht: Kluwer.

Popkewitz, T., & Brennan, M. (Eds.) (1998). *Foucault's challenge. Discourse, knowledge and power in education.* New York: Teachers College Press.

Sfard, A. (1991). On the dual nature of mathematical conceptions: Reflections on processes and objects as different sides of the same coin. *Educational Studies in Mathematics, 22,* 1–36.

Skovsmose, O. (1994). *Towards a philosophy of critical mathematics education.* Dordrecht: Kluwer.

Skovsmose, O., & Valero, P. (2001). Breaking political neutrality. The critical engagement of mathematics education with democracy. In B. Atweh, H. Forgasz, & B. Nebres (Eds.), *Sociocultural research on mathematics education: An international perspective* (pp. 37–56). Mahwah, NJ: Lawrence Erlbaum Associates.

Skovsmose, O., & Valero, P. (2002). Democratic access to powerful mathematical ideas. In L. D. English (Ed.), *Handbook of international research in mathematics education: Directions for the 21st century* (pp. 383–407). Mahwah, NJ: Lawrence Erlbaum Associates.

Valero, P. (2002). *Reform, democracy, and secondary school mathematics.* Ph.D. dissertation, Copenhagen (Denmark), The Danish University of Education.

Vithal, R., & Valero, P. (2003). Researching in situations of social and political conflict. In A.J. Bishop, M.A. Clements, F.K.S. Leung, C. Keitel, & J. Kilpatrick (Eds.), *Second international handbook of mathematics education.* Dordrecht: Kluwer.

Voigt, J. (1996). Negotiation of mathematical meaning in classroom processes: Social interaction and learning mathematics. In L. Steffe, P.Nesher, P. Cobb, G. Goldin, & B. Greer (Eds.), *Theories of mathematical learning* (pp. 21–50). Mahwah, NJ: Lawrence Erlbaum.

von Glasersfeld, E. (1995). *Radical constructivism: A way of knowing and learning.* London: Falmer Press.

Zevenbergen, R. (2000). "Cracking the code" of mathematics classrooms: School success as a function of linguistic, social and cultural background. In J. Boaler (Ed.), *Multiple perspectives on mathematics teaching and learning* (p. 201–224). Westport, CT: Ablex Publishing.

CHAPTER 4

TOWARD A POSTMODERN ETHICS OF MATHEMATICS EDUCATION

Jim Neyland

ABSTRACT

The modernist ethical orientation, evident in recent "reforms" in mathematics education, is based on the assumption that ethical self-regulation requires an ethical-legal code that redeems the pre-ethical self from a prior and unwanted disposition. Levinas provides a postmodern alternative. He argues that the ethical self is prior to all codification, and instead founded on the direct face-to-face ethical encounter of responsibility between persons. Modernity's ethical-legal code tends to erode this primary ethical orientation, with damaging consequences. In contrast to currently dominant approaches, mathematics education should ensure that legislative protocols do not override the ethical primacy of the direct encounter.

INTRODUCTION

A postmodern orientation to ethics entails a rejection, not of the ethical problems that motivate modern thinkers, but of the modern approach to

Mathematics Education Within the Postmodern, pages 55–73
Copyright © 2004 by Information Age Publishing
All rights of reproduction in any form reserved.

dealing with them. This approach is based on the assumption that ethical self regulation results from the formulation, by those in authority, of an ethical-legal code that people find reasonable to follow. In a similar way, a postmodern orientation in mathematics education does not involve a rejection of the modern concern that teachers are guided toward goals of practice, or that the profession achieves high standards. It is based on the conviction that the modern way of attempting to achieve these ends is both undesirable and illusory.

The modern way in education is exemplified in the current movement toward the scientific management of education through legislation—a movement that is behind recent education reforms. The following two examples illustrate what is happening in a number of countries and a range of other subject areas. During 1989 mathematics teachers in England and Wales received a radically new type of mathematics curriculum—one based on a conception that Skilbeck earlier identified as the technocratic-bureaucratic ideology in education (Skilbeck, 1976). This new mathematics curriculum presented school mathematics as 296 statements of attainment in 14 parallel hierarchies, most organized into ten levels. Shortly after this, the Australian Curriculum Corporation prepared booklets for mathematics teachers "containing almost 200 outcomes and well over 1000 behaviorist pointers listed in columns" (Ellerton & Clements, 1994, p. 335). From a postmodern perspective these contemporary modernist reforms are contributing to a growing malaise in education.

Mathematics enters the picture as the "paradigm case subject." That is to say that mathematics is the curriculum subject that can be used more persuasively than any other to argue that modern approaches are ill-conceived. Further, arguments based on mathematics as the paradigm case can now be supplemented by results from recent empirical studies into the personal impact of the reforms (Ball, 1997; Gewirtz, 1997; Helsby, 1999; Menter, Muschamp, Nicholls, & Ozga, 1997; Reay, 1998; Thrupp, Harold, Mansell, & Hawksworth, 2000; Woods, Jeffrey, Troman, & Boyle, 1997; Wylie, 1999). These studies suggest that the reforms may be in some way personally damaging for people who work in education. In this chapter I will investigate these modernist reforms by drawing together these arguments into a theoretical framework derived from Emmanuel Levinas' postmodern ethical philosophy.

Levinas provides a postmodern alternative to the modern way through his critique of the modern characterization of the ethical self. The modernist orientation, he argues, is based on the incorrect assumption that the ethical self is *caused* by—is the product of—social legislation that redeems the pre-ethical self from a prior and unwanted disposition. He goes on to argue that, in contrast to this, the ethical self is in fact *prior to* all ethical and rationalistic codification, and is manifest *first* in the direct relation of responsibility between individual persons. Modernity's ethical-legal code at best paralyzes the ethical self, and at worst erodes it. However, although the direct face-to-face ethical encounter of responsibility is the foundation of the ethical self, it is in itself insufficient as a substitute for an ethical code in the wider social context. Accordingly, the primary orientation of the ethical self must be supplemented, not by a modern uniform code, but by shared ethical ideals, priorities and principles that are subject to ongoing and "agonistic" negotiation. From this it can be concluded that, from a postmodern ethical orientation, mathematics education should prioritize the negotiation of ethical ideals and principles while at the same time taking steps to ensure that these *complement*, and *do not override*, the ethical primacy of the direct encounter between persons.

THE PROJECT OF MODERNITY

The "project of modernity," Habermas (quoted in Harvey, 1990) wrote, amounted to the endeavor of Enlightenment thinkers "to develop objective science, universal morality and law, and autonomous art according to their inner logic" (p. 12). They were possessed "of the extravagant expectation that the arts and sciences would promote not only the control of natural forces but also understanding of the world and of the self, moral progress, the justice of institutions and even the happiness of human beings" (pp. 12–13). Bacon, for instance, envisaged an elect group of ethicists living remote from the community and exercising control over it. Rousseau (cited in Curtis, 1981) wrote that:

> In order to discover the rules of society best suited to nations, a superior intelligence beholding all the passions of [people] without experiencing any of them would be needed.... The legislator is the engineer who invents the machine.... He who dares to undertake the making of a people's institutions ought to feel himself capable, so to speak, of changing human nature, of transforming each individual ... of altering man's constitution for the purposes of strengthening it; and substituting a partial and moral existence for the physical and independent existence nature has conferred on us all. (pp. 25–26)

He also believed that humankind would have to be forced to be free. Bentham thought that people would require the threat of coercion before they would act altruistically. Their overall goal, according to Weber, was to replace diversity with uniformity and ambivalence with order.

This modern vision is essentially monistic and entails the conception that pluralism is a precarious condition—a social disease. Monism is characterized by four assumptions. First, to all genuine questions there is one and only one true answer. Second, there are methods—different for each sphere of inquiry—which lead to correct answers in the natural sciences and in the moral, social and political realm. These true answers cannot be incompatible one with another and together form a harmonious whole. Third, even if we have not yet been able to find the appropriate method, the answers do exist. Fourth, once these solutions are discovered, reason will lead people to adopt them (Berlin, 1998a). This Enlightenment monism remains evident in contemporary technocratically oriented administration (Berlin, 1998b). In the ethical sphere it results in the attempt to bring human conduct within the domain of reason through the design of a universal ethical-legal code. It is held that because this code is both reasonable and sponsored by the state, individuals would freely choose to comply with it. Behind this is the deeper-level assumption that, without rationally designed and power-assisted rules, people would normally choose to do what is incorrect. Individual responsibility becomes, when understood in this way, the responsibility for either following or breaching the established code.

The project of modernity has been fraught with difficulties. Catastrophic social events during the twentieth century cast doubt upon the soundness of the Enlightenment vision. Contemporary human experience involves both a growing pluralism of authority and the centrality of autonomous choice in the self-constitution of individuals (Bauman, 1992; Taylor, 1991). People have shown a high degree of autonomous ethical resilience that has resulted in a tendency for them to rebel against imposed rules. And ethical codes have proven to be unable to cover the gamut of choices faced by individuals; indeed, in many respects, instead of reducing ambiguity and contradiction, they add to it. Further, modernity has contributed little to moral problematics particularly in the domain of one-to-one responsibility, and the modern way has in many instances amounted to the sanctioning of a form of moral parochialism (Bauman, 1993). Faced with these problems postmodern ethical thinkers question the wisdom of continuing the unbridled pursuit of the modern agenda. Some ask us to investigate the possibility that ethical responsibility is not primarily a *product* of society. Instead, they suggest that ethical responsibility, being *for* the other person—as distinct from being merely *with* the other person as in the modern conception of society—is the first, the *primordial*, reality of self.

THE MODERN AGENDA IN MATHEMATICS EDUCATION

Mathematics educators began experimenting with technocratic instrumentally-rationalistic approaches during the first half of the twentieth century. "Efficiency experts" (management consultants), encouraged by the way

Fredrick Taylor's Scientific Management Theory transformed the processes of industrial assembly, sought ways of applying these principles to education. Mathematics proved ideally suited for their purposes. While at first this movement toward the scientific management of education was largely confined to experimental projects, this changed in the United States during the decade following 1969. During this period a series of state legislative acts was passed making scientific management approaches mandatory. Thus the principles and procedures of scientific management became the substance of new power assisted ethical-legal codes.[1] Subsequently large scale education reform, based on these same ideas, has occurred in a number of other industrialized countries (Fiske & Ladd, 2000; Levin, 1997; Smyth, 1995; Smyth & Dow, 1998).

Four things are required when education is managed scientifically through legislation: (1) an unambiguous statement of what the legislation requires of teachers; (2) a theory of control that gives legislators confidence that teachers will do what the legislation requires; (3) agencies that will monitor the degree of compliance with legislative demands; and (4) instrumentally oriented research outputs for aiding rationalistic decision making. The consequences of these four requirements are now familiar: (1) the outcomes-based curriculum; (2) a host of quasi-legal principles, procedures and standards, and the application of contract theory to the management of professional responsibility; (3) a rise in the status of audit agencies; and (4) the unprecedented use of external consultants. The net result has been that increasingly teachers are finding their professional choices limited and their actions mediated by rules and protocols. In other words the work of teachers has steadily become more codified. This is not to say that teachers are unable creatively to interpret the various rules and protocols that are imposed upon them—only that those who do so are swimming against a tide that requires standardization. In addition, there is a related phenomenon that reinforces the effect of codification. In order to make creative choices, teachers need some sense of what is of greater value, and worthy of esteem, in mathematics education. Codification, in a single move, achieves a double feat: it strengthens minor education goals, and simultaneously weakens those major education goals that facilitate autonomous choice.

From a postmodern perspective, this loss of professional-ethical autonomy has contributed to the learning of mathematics being robbed of its enchantment. The enchanting quality of depth and mystery inspires an attitude of wonder. This attitude of wonder is more than an agreeable emotional state; it also leads, it has been argued, to learners developing as selves (Weinstein, 1975). This is a theme that is stressed by both Bauman and Varela. The former identifies, as a goal of postmodern thought, the recovery of enchantment. He describes the project of modernity as a war against mystery and magic.... At stake in the war was the right to initiative and the authorship of action, the right to pronounce on meanings, to construe narratives.... The dis-enchantment of the world was the ideology of its subordination; simultaneously a declaration of intent to make the world docile to those who would have won the right to will, and a legitimation of practices guided solely by that will as the uncontested standard of propriety.... [It] is against such a disenchanted world that the postmodern re-enchantment is aimed. (Bauman, 1992, pp. x–xi)

In his book *Ethical Know-How* Varela expresses a similar concern. He writes: "My presentation [in this book] is more than anything, a plea for a re-enchantment of wisdom, understood as non-intentional action" (Varela, 1999, p. 64).

A POSTMODERN RE-ENCHANTMENT
OF MATHEMATICS EDUCATION

A postmodern ethical approach to mathematics education, then, aims to re-enchant the world (including mathematics) that modernity has disenchanted. Postmodern re-enchantment has three components. Initially it involves establishing that the modern vision for managing and shaping mathematics teaching is neither attainable nor desirable. Supra-individual criteria for educationally sound action—in the form of technical procedures—are shown to be, at best, necessarily insufficient, and at worst damaging. Thus it is argued that a contradiction free, objectively founded and universal ethical code is impossible to obtain. It is further argued that the assertion that such a code ought to be followed, results in a weakening of the ethical self. This is because the ethical self becomes incapacitated when the essential diversity of ethical responsibility is transformed into a monotonous uniformity.

Second, postmodern re-enchantment involves reclaiming the ethical self as the center of responsible action. Ethical codes are at best a poor substitute for the direct existential relation of responsibility that is the true basis of the ethical self. It is emphasized that a postmodern orientation does not entail a disregard for modern concerns for educational well-being, only for the modern way of dealing with them. A postmodern ethics shifts the primary focus of morality away from answering to the demands of an ethical code and onto answering to either the demand of the other person who needs me, or to that of my own moral self-consciousness. Ethical codes, according to the postmodern perspective, leave out what is properly ethical in ethics. The learnable knowledge of rules is substituted for the true ethical self that is properly constituted by a relationship of direct responsibility for the other person. Human beings are neither essentially bad nor essentially good; they are essentially ethically ambivalent. That is, the existential condition of one person face-to-face with another is of ethical ambivalence. The primary ethical domain is not monotonous, regular or predictable; it is shot through with uncertainty and contradiction and cannot avoid ambiguity. A non-ambivalent ethics is thus an existential impossibility. In fact, contrary to the modernist ideal, the ethical impulse is inherently non-rational. This is because, to be ethical, the ethical impulse must precede any calculation of gains and losses or any instrumental matching of means with ends.

In the modern condition, where actions are guided by ethical codes, ethics becomes proceduralism. The only question I need to ask is, have I followed the rule? In this situation responsibility is depersonalized because it rests with my professional *role* rather than with *me* as an ethical self. Thus one of the characteristics of the modern condition is that responsibility loses its anchor with the individual person and becomes free-floating and dissipated (Bauman, 1993). The depersonalization of responsibility loops back to the earlier proceduralism establishing an inner logical consistency, because responsibility has now become a matter of following technical procedures. "Bureaucracy's double feat," writes Bauman, "is the moralization of technology, coupled with the denial of the moral significance of non-technological issues. It is the technology of action, not its substance, which is subject to assessment as good or bad, proper or improper, right or wrong" (Bauman, 1992, p. 160).

For postmodern thinkers, ethical guidelines are seen as necessary. But they do not emanate from a form of monistic legislative authority. They arise from considerations that give primacy to personal ethical responsibility via a process of ethical negotiation within a contradiction-filled ethical landscape. This negotiation is therefore "agonistic" in nature because participants are required to agonize over how contradictory ideals can best be balanced. This is exactly the opposite of the modern conception which conceives personal ethical responsibility as a *consequence* of society's rule-prescribing function. The process of postmodern ethical negotiation requires, not knowledge of prescribed procedures, standards and protocols, but openness to the freedom of the direct ethical encounter and wisdom in the face of the world's complexity and enchantment. In some ways the modernist agenda even works against itself. This is because, in its removal of the need for agonistic choice in the face of existential ambivalence, it overcomes the most modern of the modern person's attributes: the capacity for autonomous choice. Thus, paradoxically, the codification of behavior de-modernizes those in education. To put it differently, in subduing ethical responsibility, the mediation of action makes ethical agency all the more difficult.

Finally, postmodern re-enchantment involves the recovery of a relation of subjectivity between learners and mathematics. I will not pursue this component here except to note that this notion has affinities with aspects of the educational philosophies of Buber (Weinstein, 1975), Maritain (1943) and Croce (1922), and with some of Varela's (1999) recent conclusions following his studies into the science of mind, in particular with his concept of "ethical know-how" and its relationship with self.

THE ROLE OF MATHEMATICS AS
THE PARADIGM CASE SUBJECT

Mathematics education has a unique and vital role to play in fulfilment of the postmodern ethical agenda. I have called mathematics the "paradigm case subject" because it is the curriculum subject that can be used to make the strongest case against the project of modernity in education more generally. It can be used to argue that the scientific managers' goals are illusory; that by pursuing them education has been led into a cul-de-sac from which we ought to remove ourselves. If educational activity can be codified, it is the teaching of mathematics that will be most amenable to this. If any subject in general education can be represented hierarchically as a sequence of objectives for pedagogical purposes, mathematics can. From this it follows that if mathematics educators can show that mathematics *cannot* be thus represented without unacceptable anti-educational consequences, *no subject in general education can be managed scientifically.* Many of these arguments have been made by other mathematics educators and I have outlined them elsewhere, for instance, in Neyland (2001).

I will summarize these arguments briefly. At the macro-level the reform process has been criticized for the hostile nature of the discourse that accompanied it; with the anti-democratic processes that characterized its introduction; with the assumptions that underlie the epistemological, ontological and social theories used to justify these approaches; with the over-extensive use of hierarchies to organize and teach mathematics; with the ineffectiveness of outcomes approaches for teaching mathematics; and with the idea that standards of achievement can be precisely defined. At the micro-level recent modernist reforms have been challenged because they have resulted in teachers increasingly being represented as objects rather than as subjects in policy discourse; professionalism has been replaced by accountability, and collegiality by competition and surveillance; initiative, creativity and teacher-led innovation have been constrained; teaching has become technicized, and learning experiences impoverished; the workforce is overworked, distrustful, and alienated; and teachers respond pragmatically to the excessive demands of the reform mechanisms, neglecting what they believe to be the more important components of their work.

Because mathematics is the paradigm case subject, mathematics educators have a responsibility, on behalf of education more generally, to articulate fully the modern disenchantment of mathematics and its teaching and the consequent encroachment of instrumentally-rational processes into the mediation of the work of mathematics teachers. Or, to put it the other way around, mathematics educators have the urgent task of undertaking a postmodern re-enchantment of mathematics and a postmodern restoration of the primacy of the direct relationship of responsibility between teachers and students.

THE POSTMODERN ETHICAL PHILOSOPHY OF EMMANUEL LEVINAS

Levinas, one of the twentieth century's leading ethical philosophers,[2] published his comprehensive analysis of the ethical dimension of human interaction in, among other works, *Totality and Infinity* (1969), *Time and the Other* (1987), and *Otherwise than Being* (1998). His ideas have been taken up by leading scholars such as Ricoeur (1992), Sennett (1998), and Bauman (1989a, 1993).

The ethical self, viewed in the modern way, is caused by something that is prior to it, or results from either the use or rejection of something that is prior. However, this conception is problematic because, if the ethical self is to be truly ethical, it cannot be determined by some prior factor. If the ethical self is simply *caused*, then there must be something that precedes it, some foundation or determining factor—for instance, a rule, a belief in some precept, the coercive force of authority, or some physicochemical phenomenon in the brain—that makes the self ethical. Alternatively, if the ethical self is *not caused*, but something is nonetheless *prior* to it, that something could be, for instance, my rational choice that being ethical is in my best interests. This places the application of reason as prior to the ethical self. Again, if I freely choose to relinquish my self-interest in order to be ethical, then both my self-interest and reason are prior—the latter, because such a choice is made on some reasoned basis.

But what if, as Levinas argues, the ethical self is *not caused* and *nothing is prior*? Typically we find this notion difficult to conceive. This is largely because our thinking is dominated by the aetiological myth. This myth is that ethical behavior is caused by modern society's social institutions, rules, norms, and educative processes. Bauman (1989b) describes this myth as "the morally elevating story of humanity emerging from presocial barbarity" (p. 12). It is supported in many influential social theories, and contrary theories have "a long way to go before they succeed in displacing [this] myth from public consciousness" (p.12). Lay opinion, on the whole,

> resents all challenge to the myth, [and this] resistance is backed ... by a broad coalition of respectable learned opinions which contains such powerful authorities as the "Whig view" of history as the victorious struggle between reason and superstition; Weber's vision of rationalization as a movement toward achieving more for less effort; [the] psychoanalytical promise to debunk, prise off and tame the animal in [human persons]; Marx's grand prophecy of life and history coming under full control of the human species once it is freed from the present debilitating parochialities; Elias's portrayal of recent history as that of eliminating violence from daily life; and, above all, the chorus of experts who assure us that human problems are matters of wrong policies, and that right policies mean elimination of problems. Behind the alliance stands fast the modern "gardening" state, viewing the society it rules as an object of designing, cultivating and weed-poisoning. (Bauman, 1989b, pp. 12–13)

According to this viewpoint, society causes ethical behavior, and unethical behavior is evidence of failure within society's institutions. This view grounds much orthodox social and educational theory, and it is contrary to Levinas' postmodern ethical orientation.

I have argued elsewhere (Neyland, 2001) that the aetiological myth can be challenged on the grounds that it is based on two unsustainable premises: (1) that modern societies demonstrate a level of reflexivity not present in pre-modern forms of life; and (2) that our social practices are structures of rules to be followed by self-cognizing agents, with these rules having a qualitatively inferior form in pre-modern life.

Reflexivity entails the monitoring of both actions (one's own and those of others), and shared knowledge constructs. Such monitoring is rarely fully conscious. Some social theorists—Garfinkel, Goffman, Giddens and Barnes, for instance—have cast significant doubt on the notion that modern life is characterized by relatively higher levels of reflexivity. Giddens (1991) argues that reflexivity is "continuous" and "all pervasive" in social situations (pp. 75–77); Garfinkel (1967) maintains that high levels of reflexivity are essential to social life of all forms; Goffman (see Giddens, 1990) claims that a great deal of human action involves a consistent and never-to-be-relaxed monitoring of behavior and its contexts; and Barnes (1995) insists that all societies, modern and pre-modern, exhibit an essential openness through reflexivity without which they would cease to be societies.

If social practices are structures of rules, these rules need to be internalized by people, so that they can subsequently act in accordance with them. However, Barnes points out that if rules are internalized they become finitely stored information. But this cannot happen because they need to be able to be exercised in indefinitely varying and complex situations. Instead, agents act on the basis, not of rules, but of previous actions through analogical extensions of them, and these in turn shape how others determine what counts as a rule. Thus people are required to reconstruct shared understandings of what is implied by a norm and this renders it public, not private and internalized; it exists as agreements in practice. In fact, social order is dependent upon norms not being internalized (Barnes, 1995). MacIntyre (quoted in Bernstein, 1983) makes a similar argument. He writes:

> Objective rationality is ... to be found not in rule-following, but in rule-transcending, in knowing how and when to put rules and principles to work and when not to. Because there is no set of rules specifying necessary and sufficient conditions for large areas of such practices, the skills of practical reasoning are communicated only partly by precepts but much more by case-histories and precedents. Moreover the precepts cannot be understood except in terms of their application in the case histories; and the development of the precepts cannot be understood except in terms of the history of both precepts and case histories. (p. 57)

According to Levinas, and again in contradistinction to the aetiological myth, ethical responsibility precedes all engagement with the other, and has no foundation or determining factor; it is not caused and nothing is prior to it. There is no self before the ethical self. Ethics precedes being; or, to use Levinas' expression, it is "otherwise than being" (Levinas, 1998). It is in the realm-not-realm that is *better* than being. It is better because an ethical self that is based on calculations of gain, or that entails a denial of a prior self-interested self, is not much of an ethical self. What, then, constitutes the ethical self? It is based on the experience of an encounter with another person, face-to-face.

There are some affinities in this respect between Levinas' thought and the distinction the education philosopher Martin Buber made between the *I-It* and the *I-Thou* relationship between people.[3] For Buber, the world is not only the world of objects but also the world of relations, such as the person to nature and one person to another. In the *I-It* relation, both physical objects and other people are viewed instrumentally, and other people are conceived in what Jeffreys (1957) calls "dossier terms" (p. 66). Buber (quoted in Curtis & Boultwood, 1966) maintained that it is essential for each of us to rediscover the *I-Thou* relationship. This relation is unmediated. That is,

> No system of ideas, no foreknowledge, and no fancy intervene between the I and Thou. The memory itself is transformed, as it plunges out of its isolation

into the unity of the whole. No aim, no lust, and no anticipation intervene between I and Thou. Desire itself is transformed as it plunges out of its dream into the appearance. Every means is an obstacle. Only when every means has collapsed does the meeting come about. (p. 611)

The *I-Thou* relationship, Buber (1958) argues, is one of love, sympathy and respect. However, the *I-Thou* relationship cannot last forever, even between close friends. We are obliged to live in a world of objects; we cannot survive as a society without this relation. But it is in the world of the *I-Thou* that we discover meaning for our lives, as people and as educators.

Like Buber, Levinas draws attention to a deeper quality in the relationship between people. This is evident when people communicate. Immanent in every "said," he writes, there is first a "saying." A discourse is said and usually addressed to another. Speaking is at base communication. The "said" presumes a relation between a speaker and a hearer. The person to whom I speak is not primarily there before me as an object or phenomenon that I observe and analyze, and my relation to her is not primarily intentional. She is someone to whom I make an offering (Peperzak, 1989). The "saying" and the "said" occupy two different dimensions—the ethical and the ontological—that, while being intimately related, cannot be synthesized into a totality. The "saying" bears a "said" in the same way that a society needs laws and institutions. But the saying is in essence a response to the "face" of the Other. There is an anarchical quality about the saying that is not reducible to the said; the relationship between people is the paradigm example of a non-synthesizable meaning. In the words of Levinas (1985): "The true union of true togetherness is not a togetherness of synthesis, but a togetherness of face-to-face" (p. 77). When I speak to another person I am first acknowledging her as another person. When I focus on her "face" I do more than just gaze; I actually encounter her. This is where Levinas' thought begins to diverge from Buber's. This encounter, Levinas argues, is at its deepest level, *an awareness of the other as one who needs me.* This experience of a call to responsibility is the source of the social bond. This encounter is primarily ethical and forms the basis of the ethical self.

In a more pronounced departure from Buber, Levinas insists that the relationship between people is *not symmetrical* or reciprocal, as in Buber's thought, but necessarily *asymmetrical.* My responsibility is not dependent upon the Other reciprocating. That is the other's affair (Levinas, 1985). Being *with* the Other is symmetrical; being *for* her is asymmetrical. Unlike Sartre, who sometimes interpreted the relationship with the Other as a threat, Levinas argues that the primordial call to responsibility for the Other is a call to selfhood. "The face of the Other is destitute; it is the poor for whom I can do all and to whom I owe all" (Levinas, 1985, p. 89). In my response to the Other, I find myself. Responsibility is constitutive of the self. It is not an attribute assumed by a prior free and autonomous self; it does not augment subjectivity. Subjectivity and responsibility come at the same moment. In other words, subjectivity is ethical. The latter is "the essential, primary and fundamental structure of subjectivity ... the very node of the subjective is knotted in ethics understood as responsibility" (Levinas, 1985, p. 95). I am a self because I am exposed to the Other. I become a responsible self to the extent that I relinquish my ego-centrality and open myself to the vulnerable other. Of course, the ethical demand of the face-to-face is not a causal necessity. I can treat others badly.

The ethical "I," then, not being founded upon the assumption of reciprocity, is not a singular version of the plural "we', and not interchangeable with "he" or "she" (Bauman, 1993). I am for the Other irrespective of whether the Other is for me or not. As Bauman observes, paradoxically, if there is a rule it is a *singular* rule for me only. Again, as I observed earlier, if this were not the case, the self would not amount to much ethically. However, the Other's command of me does not lead to my humiliation. I command the Other to command me (Bauman, 1993). Thus, Levinas' notion is fundamentally different from the social "contractual" situation of modern thought where duties and sanctions are negotiated in advance. Duties, Bauman argues, make people alike. Responsibility makes people individuals. "Humanity is not captured in common denominators—it sinks and vanishes there. The morality of the moral subject does not, therefore, have the character of a rule" (p. 54). That is, the ethical dimension resists codification.

> Being moral means being abandoned to my own freedom... I am moral *before* I think. There is no thinking without concepts (always general), standards (general again), rules (always potentially generalizable). But when concepts, standards and rules enter the stage, moral impulse makes an exit; [a codified ethics] is made in the likeness of Law, not the moral urge. (Bauman, 1993, pp. 60–61)

Society, needless to say, is composed of more people than just myself and the immediate other whom I face. Accordingly, I am required to moderate and extend the primary relationship to include others. The ethical dimension must be complemented by the ontological dimension that involves totalizing concepts and the compression of the ethical dimension to one of simultaneity with the ontological. This is where justice functions through social institutions and conventions. The ethical relation of the face-to-face must become political and juridical through a societal attempt to achieve the most important mode of simultaneity—the simultaneity of equality. Thus, while the ethical relation would suffice for the immediacy of the face-to-face, it is necessary to attempt to define the indefinable others in order to try to balance the conflicting claims found within the community, and to make the sorts of comparison demanded by the obligation to administer a just society. But, and this is critical, *these social institutions must always be held in check by the primacy of the initial interpersonal relationship.* Levinas opposes the dominant view that social institutions have a vital role in suppressing the capricious, unrefined and barbaric in human nature. He acknowledges that social institutions and organizations have a role in weakening individual delusions, but he argues that the orthodox understanding leads to forms of social suppression that are damaging to human interrelations.

> The demands of justice arise from out of the ethical situation and at the same time pose a danger for that situation. The danger of justice, injustice, is the forgetting of the human face. The human face "regulates," it is the goodness of justice itself. (Cohen, 1986, p. 9)

I now turn to the central question for this chapter: What causes the erosion of the ethical relation and hence damages us as selves? The relation of responsibility is maintained by what Levinas calls ethical *proximity.* Proximity is the realm of intimacy and morality. It is not a spatial, cognitive or emotional closeness; it exists insofar as I feel or am responsible for the other person. Modern thinkers see individuals as instrumentally related and perceive between them a distance structured solely by legal codes. Levinas' postmodern ethics returns us to the ethical significance of proximity. This is what make his ethics postmodern (Bauman, 1993). However, the state of proximity is entangled in contradiction without solution. This is because we cannot escape the fundamental human condition of being.

> Since it is in the state of proximity that the responsibility, being unlimited, is least endurable—it is also in the state of proximity that the impulse to escape responsibility is at its strongest .... [It] is the terrain of morality's most dazzling glory and its most ignoble defeats. (Bauman, 1993, pp. 88–89)

Ethical proximity can be undermined and accordingly ethical selves can be weakened. Kelman (1973) argues that ethical relations are eroded when three conditions are met: procedures are authorized, actions routinized, and people dehumanized. The first two are in many respects goals of the scientific management of mathematics education. The last is a consequence of it. Mathematics students are increasingly represented in "dossier" terms and thus dehumanized. And mathematics teachers are viewed instrumentally by those in authority, and perhaps increasingly by their students.

CONCLUSION

The application of a postmodern ethical orientation to mathematics education will not make teaching less complicated, or the administration of education smoother. But it will make it better. It will be better because it will shift the focus away from procedural compliance and onto the direct ethical relationship between mathematics teachers and their students. From this perspective, all collective guidelines for the teaching of mathematics must first be passed through the sieve of the ethical. Teachers must also be given space to form such collective agreements through a process that involves the articulation of ideals and the analysis of their consequences. Because ideals typically conflict with each other, teachers will also need the opportunity to "train their imaginations to go visiting," to use one of Hannah Arendt's expressions. That is, they will have to learn, through the exercise of imagination, to appreciate—if not espouse—something of those ideals esteemed by colleagues; especially those that grate. In addition, teachers will require the space to work "agonistically" toward a balance between these various and conflicting ideals.

This process will necessarily be ongoing because any agreements will be localized and provisional. If the perspective is fully postmodern, some of these ideals will address mathematics as something that is enchanting, worthy of our esteem, and evocative of wonder. The postmodern vision for mathematics education is motivated by ideals. But these ideals are not utopian. They are plural and contradictory and exist in a dynamic relation that requires an ongoing effort of imagination, a continual renegotiation of priorities, and a steadfast vigilance against a modernist disposition toward totalizing and control.

A postmodern re-enchantment is not a panacea. But what are the implications of continuing, as at present, with the modernist scientific management of education? Earlier in this chapter I cited evidence that the codification of action is leading to damaged teachers; to what might be called, a loss of spirit. What is happening to mathematics itself? The acclaimed mathematics educator, Hans Freudenthal (1978), clearly identifies the consequences for mathematics. The "image of mathematics [currently presented is one] which every mathematician will detest"; it is "a caricature of mathematics"; "it is a wrong picture of mathematics" (pp. 96–98). It hardly needs saying that mathematics teaching based on such an image of mathematics is not good for learning.

What would be the consequence of a postmodern re-enchantment of mathematics education and the adoption of a postmodern ethical orientation? This is a matter for speculation—but it is clear where we ought to begin if we wish to move in this direction. I will suggest three starting points: the negotiation of shared ideals, ideas-based mathematics, and learning situations that encourage surprise and joy in the learner. "Education," wrote Weil (1952), "consists in creating motives" (pp. 181–182). This means, for both teachers and students, restoring to their activities a sense of purpose and creative spontaneity. This requires a shared sense of what is worthy of esteem in mathematics and its teaching and learning.

Associated with this is the requirement that mathematics be presented, not as hierarchical, logically consistent, and rule-based, as at present, but as ideas-based (ideals are discounted in hierarchical and rule-based mathematics, but play a prominent role in ideas-based mathematics). Ideas-based mathematics is essentially humanistic—as distinct from formalistic—because ideas are human inventions. Mathematical ideas reside in the dialogical space between structure and creativity; between proof and refutation. For such an understanding of mathematics there is no better place to start than a Lakatosian quasi-empirical philosophy of mathematics (Lakatos, 1976).

Freudenthal (1978) argues that we ought to abandon hierarchical and formalistic approaches and focus mathematics education around open learning situations. "The most fruitful result of an open learning situation," he writes, is "that it makes surprises possible" (p. 180). Mathematical surprise? This is, he argues, "what matters most" in mathematical learning, and he calls this—in an attempt to underline its central importance—his "first thesis" (p. 165). He wrote this at the time when the consequences of the scientific management of mathematics education were becoming evident in the United States. Surprise is the thing that matters most! This is clearly a call for a recovery of enchantment. In the book, *Better Mathematics* (Ahmed, 1987)—which represents an approach to professional development at the opposite end of the spectrum from that associated with the scientific management of education—surprise and joy are given central importance along with a small number of other qualities that together resonate with the theme of re-enchantment and postmodern ethics. *Better Mathematics* outlines the beginning of a conversation that we might collectively continue. The conversation is, in effect, about three things: re-enchantment, ideas-based mathematics, and the negotiation of ideals. It is, in other words, the beginning of a conversation that would lead to a re-conception of mathematics education in postmodern ethical terms.

NOTES

1. This began with the United States Federal Government's National Assessment of Educational Progress project in 1969. A raft of state level activity followed from this. For instance, in 1971 California appointed an Educational Management and Evaluation Commission. Nine public members were to be appointed, three to represent the field of economics, three to

represent the management sciences, and three to represent the learning sciences. In the same year, another state, Colorado, passed the Educational Accountability Act in which reference was made to the legislative tools of scientific management in education.

2.   See, for instance, Nemo's introduction to Levinas (1985).

3.   The educational philosopher Martin Buber wrote about this: "The 'I' of the primary word 'I-Thou' is a different 'I' from that of the primary word 'I-it.' The 'I' of the primary word 'I-it' makes its appearance as an individuality and becomes conscious of itself as *subject* (of experiencing and using). The 'I' of the primary word 'I-Thou' makes its appearance as person and becomes conscious of itself as person and becomes conscious of itself as *subjectivity* (without a dependent genetive)" (cited in Weinstein, 1975, p. 26, emphasis added).

REFERENCES

Ahmed, A. (Ed.). (1987). *Better mathematics: A curriculum development study.* London: Her Majesty's Stationery Office.

Ball, S. (1997). Good school/bad school: Paradox and fabrication. *British Journal of Sociology of Education, 18*(3), 317–336.

Barnes, B. (1995). *The elements of social theory.* London: University College London Press.

Bauman, Z. (1989a). Social manipulation of morality: Moralizing actors, adiophorizing action. In Z. Bauman (Ed.), *Modernity and the holocaust* (pp. 208–221). New York: Cornell University Press.

Bauman, Z. (1989b). *Modernity and the holocaust.* New York: Cornell University Press.

Bauman, Z. (1992). *Intimations of postmodernity.* London: Routledge.

Bauman, Z. (1993). *Postmodern ethics.* Cambridge, MA: Blackwell Publishers.

Berlin, I. (1998a). The apotheosis of the romantic will. In H. Hardy & R. Hausheer (Eds.), *The proper study of mankind: An anthology of essays* (pp. 553–580). New York: Farrar, Straus & Giroux.

Berlin, I. (1998b). Two conceptions of liberty. In H. Hardy & R. Hausheer (Eds.), *The proper study of mankind: An anthology of essays* (pp. 269–325). New York: Farrar, Straus & Giroux,

Bernstein, R. (1983). *Beyond objectivism and relativism: Science, hermeneutics, and praxis.* Philadelphia: University of Pennsylvania Press.

Bruber, M. (1958). *I and thou.* New York: Charles Scribner's Sons.

Cohen, R. (1986). Introduction. In R. Cohen (Ed.), *Face to face with Levinas* (pp. 1–12). Albany: State University of New York Press.

Croce, B. (1922). *Aesthetic.* New York: Macmillan.

Curtis, M. (Ed.). (1981). *The great political theories.* New York: Avon Books.

Curtis, S., & Boultwood, M. (1966). *A short history of educational ideas.* London: University Tutorial Press.

Department of Education. (1987). *Better mathematics: A curriculum development study.* London: Her Majesty's Stationery Office.

Ellerton, N., & Clements, M. (1994). *The national curriculum debacle.* Perth: Meridian Press.

Fiske, E., & Ladd, H. (2000). *When schools compete: A cautionary tale.* Washington, DC: The Brookings Institution Press.

Freudenthal, H. (1978). *Weeding and sowing: Preface to a science of mathematics education.* London: Reidel Publishing Co.

Garfinkel, H. (1967). *Studies in ethnomethodology.* Englewood Cliffs, NJ: Prentice-Hall.

Gewirtz, S. (1997). Post-welfarism and the reconstruction of teachers' work in the UK. *Journal of Education Policy, 12*(4), 217–231.

Giddens, A. (1990). *The consequences of modernity.* Stanford, CA: Stanford University Press.

Giddens, A. (1991). *Modernity and self-identity: Self and society in the late modern age.* Oxford: Polity Press.

Harvey, D. (1990). *The conditions of postmodernity.* Cambridge, MA: Blackwell Publishers.

Helsby, G. (1999). *Changing teachers' work: The reform of secondary schooling.* Buckingham: Open University Press.

Jeffreys, M. (1957). Existentialism. In A. Judges (Ed.), *Education and the philosophic mind* (pp. 60–80). London: George G. Harrap & Co.

Kelman, C. (1973). Violence without moral restraint. *Journal of Social Issues, 29,* 29–61.

Lakatos, I. (1976). *Proofs and refutations.* Cambridge: Cambridge University Press.

Levin, B. (1997). The lessons of international education reform. *Journal of Education Policy, 12*(4), 253–266.

Levinas, E. (1969). *Totality and infinity: An essay on exteriority* (Trans: A. Lingis). Pittsburgh, PA: Duquesne University Press.

Levinas, E. (1985). *Ethics and infinity: Conversations with Philippe Nemo* (Trans: R. Cohen). Pittsburgh, PA: Duquesne University Press.

Levinas, E. (1987). *Time and the other* (Trans: R. Cohen). Pittsburgh, PA: Duquesne University Press.

Levinas, E. (1998). *Otherwise than being: Or beyond essence* (Trans: A. Lingis). Pittsburgh, PA: Duquesne University Press.

MacIntyre, A. (1977). Epistemological crises, dramatic narrative and the philosophy of science. *Monist, 60,* 453–472.

Maritain, J. (1943). *Education at the crossroads.* New Haven: Yale University Press.

Menter, I., Muschamp, Y., Nicholls, P., & Ozga, J. (1997). *Work and identity in the primary school—a post-Fordist analysis.* Buckingham: Open University Press.

Neyland, J. (2001). *An ethical critique of technocratic mathematics education: Towards an ethical philosophy of mathematics education.* Unpublished PhD Thesis, Victoria University, Wellington.

Peperzak, A. (1989). From intentionality to responsibility: On Levinas's philosophy of language. In A. Dallery & C. Scott, (Eds.), *The question of the other: Essays in contemporary continental philosophy* (pp. 3–21). Albany: SUNY.

Reay, D. (1998). Micro-politics in the 1990s: Staff relationships in secondary schooling. *Journal of Education Policy, 13*(2), 179–196.

Ricoeur, P. (1992). *Oneself as another* (Trans: K. Blamey). Chicago: University of Chicago Press.

Rouseau, J-J. (1906). *The social contract* (Trans: G.D.H. Cole). London : Dent.

Sennett, R. (1998). *The corrosion of character: The personal consequences of work in the new capitalism.* New York: W. W. Norton & Co.

Skilbeck, M. (1976). Ideologies and values Unit 3 Course E203. *Curriculum design and development.* Milton Keynes: Open University.

Smyth, J. (1995). What's happening to teachers' work in Australia? *Education Review, 47,* 189–198.

Smyth, J., & Dow, A. (1998). What's wrong with outcomes? Spotter planes, action plans, and steerage of the educational workplace. *British Journal of Sociology of Education, 19*(3), 291–303.

Taylor, C. (1991). *The ethics of authenticity.* Cambridge, MA: Harvard University Press.

Thrupp, M., Harold, B., Mansell, H., & Hawksworth, L. (2000). *Mapping the cumulative impact of educational reform: A study of seven New Zealand schools.* Hamilton: University of Waikato.

Varela, F. (1999). *Ethical know-how: Action, wisdom, and cognition.* Stanford, CA: Stanford University Press.

Weil, S. (1952). *The need for roots.* London: Routledge and Kegan Paul.

Weinstein, J. (1975). *Buber and humanistic education.* New York: Philosophical Library.

Woods, P., Jeffrey, B., Troman, G., & Boyle, M. (1997). *Restructuring schools, reconstructing teachers: Responding to changes in the primary school.* Buckingham: Open University Press.

Wylie, C. (1999). *Ten years on: How schools view educational reform.* Wellington: New Zealand Council for Educational Research.

part II

POSTMODERNISM WITHIN CLASSROOM PRACTICES

Part II relates to the postmodern suspension of truth and critique of universal reason. The four chapters in Part II are set within the classroom. The production of student knowledge is the focus, and issues of agency, power and unconscious processes are considered as central to the development of a conceptual space for knowledge production.

CHAPTER 5

FACILITATING ACCESS AND AGENCY WITHIN THE DISCOURSES AND CULTURE OF BEGINNING SCHOOL

Agnes Macmillan

ABSTRACT

This chapter focuses on relationships between language and learning and their location in social contexts. In it I explore very young mathematics students' potential to gain access to the new meanings of the formal mathematics classroom as they talk about what is already known in relation to what is new; maintain the sense of individuality, creativity and agency, and natural propensities for learning they bring with them to school; accept and understand the need for accuracy, precision and the conventions of mathematics; and acquire the discursive strategies that allow them to develop an identity as a student of mathematics.

Mathematics Education Within the Postmodern, pages 77–101
Copyright © 2004 by Information Age Publishing

INTRODUCTION

In the mathematics classroom students are learning to do mathematics and learning to become mathematical. This process of "learning to do" involves the communication of mathematical ideas and participation in mathematical activities. That is, it happens within a social context with others listening and talking and involves writing and drawing as children explore, practice, problem solve, and reflect on what they are doing.

Connections made between language and learning during the past three decades have produced a wealth of new understanding about what is being generated socially and culturally—that is, interpersonally. Those connections have become possible through new theorizing about descriptions and interrogations of social practices. Postmodern theories in general, and critical discourse analysis theory in particular, provide tools for investigating how forms of social practice constitute individuals as thinking, feeling, and acting subjects. Links between language and knowing, like these, have implications for mathematics learning, and mathematics education is beginning to be conceived as having its meaning potential located in the particular social context in which it occurs (Chaiklin & Lave, 1993; Lave & Wenger, 1991; Noss, 1998; Resnick, Levine, & Teasley, 1991).

Children become literate and numerate by being immersed in language—spoken and written—by coming to know and critically view all types of literary and factual texts. The desire to be numerate emerges from shared experiences that allow for integration of known and new knowledge. Personalized conversations guide and monitor the individual toward the acquisition of new skills and strategies, modeled by readings and writings. These teaching-learning strategies create a sense of community—of acceptance and equitable access to the numerate culture's knowledge base. It is within classroom communities of cohesion and collaboration, created by teachers, that equal access to the activities and discourses of the numerate culture (Australian Association of Mathematics Teachers, 1997, 1998) is said to be created. It is the teacher who creates the conditions to support inclusiveness, confidence, competence and empowerment.

But we need to know more about the kinds of talk and kinds of activities that allow equal access to knowledge and empowerment for students. Here in this chapter I focus on classroom talk—albeit set within a context of activity—in order to discover what kinds of talk help very young students to make sense of mathematical meanings: How do children use particular forms of language to acquire knowledge of the mathematical system? As an abstract system, like literacy, it contains knowledge of what and why—conceptual knowledge—as well as about how, when and where—procedural knowledge. There are rules, beliefs and values generated discursively—that is, there are particular ways of saying and doing things. Is it possible for children to become immersed in those numerate meanings through talk and activity? Do they explain, report, discuss, argue and evaluate as they explore, practice, problem solve and reflect? If so, how easily can the early years classroom talk create equitable access to knowledge; how comfortable can the teacher be in positioning the children as their own meaning makers; how can shared meaning making be generated and how can it be monitored, guided and assessed? How does it inform what we plan to do next—or is it more important to think about setting up the children in a context that is both secure and invigorating?

These themes of discourse, as ways of social meaning construction, will be threaded through this chapter. The chapter has three main goals. First it explores a perspective on the modeling of the social and cultural context and offers a framework for looking at the different types of oral text along with the psychological strategies that stimulate cognitive engagement. The second section provides selections of transcripts from small-group mathematics activities of a first-year-at-school class. Here I deconstruct or unpack the language in order to determine how the children and teacher talk and work together, to make sense of the meanings being generated from within the activities. In the third section I discuss how well the children were able to make sense of what they were doing by participating in the activities and discourses of the numerate culture. To what extent could they blend known and new knowledge; allow natural learning potential, individuality, and creativity to be fulfilled; understand and accept the rules and values; and negotiate their own ways through the numerate meanings?

MODELING THE WAYS OF SAYING AND DOING THINGS

The critical literacy frameworks (Halliday, 1994; Luke, 1993) of current curriculum documents have offered teachers ways of understanding different types of texts in terms of their structures, modes, functions and purposes. These types of texts—logical, clarifying and procedural explanations, reports, recounts, arguments, discussions and evaluations—are being regarded as central to acquiring appropriate ways of communicating in the mathematics classroom. (Definitions of these generic structures are presented in the Appendices, Table 5.1. The functions and structures are derived from Gerot and Wignell [1994], and the summaries from Marks and Mousley [1990]). At the interpersonal level of communicating at school, messages are generated directly and indirectly about the necessity for accuracy, precision and adherence to the conventions of mathematics. By this is meant that teachers emphasize these conventional aspects of mathematics in ways that may facilitate or inhibit access to the mathematical meanings and the learners' evolving interpretations and creative potential.

INTERPRETATIONS AND CREATIVE POTENTIAL

The term "modeling" refers to the processes of perceiving, interpreting and understanding new procedures and practices, codes and mores, and incorporating these into one's consciousness. Such internal functions create impetuses for particular kinds of sociocognitive or regulatory strategies to be applied through communication. Sometimes referred to as discursive strategies, they form the rationale for examining language as social practice—for generating mental and social integration, learning to negotiate and to reproduce meaning (Bourdieu, 1977; Davies, 1993; Davies & Harré, 1990; Lave & Wenger, 1991; Lloyd & Duveen, 1993). Typical strategies used by learners are imitating, checking, testing (hypothesizing), anticipating, imitating, assisting, positioning and improvising. (Definitions of these concepts are presented in the Appendices, Table 5.2.)

Upon examining the playing discourses of informal educational settings (Macmillan, 1997), processes considered to be psychological and cognitive were found to be socially driven. Since they were unconsciously and consciously realized through talk and activity in the social context, they complement and support sociocultural modeling processes (Macmillan, 1998a, 1998b, 1998c, 1999). They are familiar to most educators as the processes linked to Piaget's accommodation and assimilation strategies and are identified here as intentionality and systematicity—adaptability, integration, transferability and evaluation (Peirce, 1955, 1966).

When talking about the children's conversations with each other and the teacher in the next section, each type of oral text is being related to one or more sociocognitive processes. That is to say, in general,

- clarifying, logical, and procedural explanations are functioning as *intending, adapting, integrating* modes;
- reports and recounts are functioning as *integrating, transferring* modes;
- arguments and discussion are functioning as *transferring* modes; and
- evaluations are functioning as *critically reflecting, awareness* modes.

The sociocultural outcomes of such processes are being accounted for as the accessibility of the meanings. The engagement, participation, and negotiation affordance are equity issues. (Definitions of these are presented in the Appendices, Table 5.3). A diagrammatic summary of these sociocultural, linguistic and sociocognitive dimensions of teaching-learning experiences and their relationships to sociocultural outcomes is provided in Figure 5.1.

An overview of the activities selected for discussion and the main analytical focuses for each of them is provided in Figure 5.2.

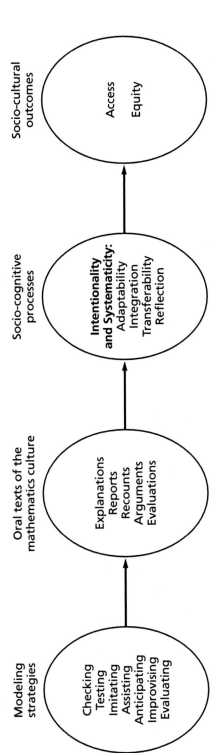

Figure 5.1. Relationships between modeling strategies, the register of mathematics, sociocognitive processes and socio-cultural outcomes

Activities	Modeling Strategies	The Register	Outcomes
The family graph activity	Reflection	Recounts, Evaluations	Co-participation
The fractions activity	Anticipating, Checking, Demonstrating	Procedural questions and statements	Responsible self-regulation
Classifying activities	Improvising	Clarifying statements and questions	Engagement
The line-matching activity	Reflection	Discussion	Negotiability

Figure 5.2. Overview of the mathematics activities and main analytical focus

BACKGROUND TO THE STUDY

In this chapter I draw on a small selection of data from an ethnographic study that was conducted in two preschools and two Kindergarten classes in Australia (see Macmillan, 1998a, for reports of the preschool data). Nonparticipant observations using field notes, tape and video recordings were carried out on a daily basis over two six-week periods at the two preschool sites. Eight children from both of the preschools were followed into their first-school classes, where participant observations using the same techniques were carried out on a daily basis over a six-month period. In the first year at school, teachers often organize part of the day as small-group, "hands-on" activity sessions. In the site being reported here the groups rotated each twenty minutes or so, and the children in the other groups were supervised by the classroom teacher and one or two parent helpers. The author as researcher became the teacher of the mathematics activity during these sessions, and this allowed her to interact with all the children in the class two or three times in the course of a week. For the duration of the warm weather, the mathematics activities were held just outside the classroom on an adjoining verandah. Toward winter, we joined the other children indoors.

TALKING WITHIN AND ABOUT THE ACTIVITIES

The Family Graph Activity

This activity required children to select from a collection of photocopied pictures of infants, children and adults, representing the members of their family and paste them beside each other on a piece of card. In particular, they were required to:

- interpret pictorial representations of people of various stage of maturity who may or may not resemble members of the children's families;
- decide which pictorial representations could most accurately represent the members of their family;
- create one-to-one correspondence between each family member and a picture; and
- cut out the pictures and arrange them in logical order from left to right on a blank piece of card, thereby creating a graphic representation of their family.

The children attempted to reconcile their conceptions of the attributes of the family members represented in the pictures with their own family members by looking for pictures that were as much like their own family members as possible. For example, children with much younger looking grandmothers than the one depicted as the very old lady in a rocking chair with glasses, found it difficult to accept the given image of a grandmother. In the following conversation the two children were adapting and integrating known conceptions about who was in their family and their familial identities and relationships with the new context.

1.   Ch M: I'm going to cut his hat off.
2.   My grandfather doesn't have a moustache. [*repeats it and laughs*]
3.   Ch A: Your brother doesn't look like that.
4.   Ch M: I know, but he has a hat like this but a kind of different color.
5.   And my "poppy" isn't like that or doesn't have a chair like that.
6.   Ch A: Yeah, 'cause he's got a moustache. Look.
7   I've never seen somebody, an old man with a moustache.
8.   Ch M: I didn't hear of my grandfather with a moustache.
9.   Ch A: Here, they look silly with moustaches.
10.   Ch M: And I do have an uncle. That might be my uncle.
11.   Hey, I got to cut around him.
12.   Ch A: Yes, this might be my uncle too.
13.   'Cause my uncle has a moustache.

As they viewed the images the children critically evaluated mismatches concerning the physical attributes of the people they knew and the people in the pictures (lines 2–10, 12–13), and they used clarifying (lines 2, 4, 7) and logical explanations (lines 6, 13) to support their case. Procedural explanations were used to report on the cutting out process (lines 1, 11).

Later, another child engaged in critical evaluation when he stated: "My grandpa doesn't wear glasses all the time, only when he reads the newspaper." As another child looked for a picture which would provide a suitable representation of his big sister, he used procedural, clarifying and logical explanations to justify his choice of picture: "I need a ... I need a big sister. So I'll put my big sister on it. But she doesn't usually wear high heels, 'cause she doesn't have high heels."

Communicating in the register of mathematics, and exploring concepts of cardinality, ordinality, temporal and physical measurement, were evident in the following conversation:

1.   Ch G: My big brother, Robert, he's bigger than Mum.
2.   T: Is he? How old is he?
3.   Ch G: He's about thirteen I think.
4.   T: Mmm. And who's between you and your brother? Anyone?

5.   Ch G: Yeah. [*Long pause*]
6.   T: You've only got one brother have you?
7.   Ch G: I've got two brothers.
8.   T: A big brother....
9.   Ch G: One's at high school.
10.   T: And that's your baby brother [being cut out].
11.   Ch G: Yes.

This child used clarifying explanations (lines 1, 3, 7, 9) to reflect on family members' ages and positions. His response about the age of the brother who was bigger than his mum ("about thirteen") signified some understanding of number concepts beyond ten; and the qualifier, "I think," provided evidence of further reflective awareness as he noted that it was his own perception and may not be accurate (line 3).

One child did not want to represent her father in the graph because he did not live with her any more and, according to her reasoning, was therefore no longer her dad. She changed her mind, however, as she recalled an event and recounted it: "We do have a dad ... But we go over, and we've got a dog and he runs away and he went to my aunty's once and the little dog was chasing him." This recount of an experience allowed her to reflect on, and, for the moment at least, adapt to the idea that her dad could still be perceived as being in a familial relationship with her.

One of the main mathematical challenges of this activity involved placing the differently shaped and sized pictures so that they would fit on the piece of card.

1.   Ch J: Where will I put the sister, and the brother?
2.   T: Beside them.
3.   Ch J: Yeah, well what about the grandma?
4.   T: I don't know, you might have to take two pieces of card.
5.   Put them as close together as you can.
6.   Do you want some help with your cutting?
7.   Ch D: Yeah.

This child asked procedural questions regarding the location of the pictures (lines 1, 3), and the teacher modeled the socially desirable strategies of assisting and supporting each other (lines 6, 7). The teacher's positioning language generated mathematical and cultural meanings: she used mathematical terms to model specificity about where the people were to be placed (lines 2, 5). The children's attention to the task's requirements was evident in discursive strategies linked to intentionality and systematicity—the choice, identification and positioning of family members were all negotiable. At the same time the teacher responded to the children's inquiries with clear explanations of the desired modes of presentation. For this task, challenges could be resolved in ways that did not threaten individual or interpersonal motivations. The literate and mathematical purposes of the task demanded processes of reflective abstraction, but individual interpretations led to negotiable frameworks and buoyant self-regulative capacities.

The Fractions Activity

For this activity lesson the children used squares and circles of colored paper to make halves by folding them. Similar procedures were undertaken for the concept of a "quarter." Following discussions and experiments with colored paper squares and circles, children were asked to fold and cut out halves and quarters, and then paste them back together as a whole circle or square. The children were asked to "write" a story on the sheet to record what they had done.

1.	Ch M: Fold it a different way. I've got a rectangle.
2.	T: Pretend it's a handkerchief. Fold it in half and half again.
3.	Ch M: I done it. I'm having a brown handkerchief.
4.	T: [*Later*] This is one half….
5.	Ch R: When are we going to stick these on?
6.	T: In a minute.
7.	Ch B: Open it?
8.	T: Open it. Yes.
9.	Ch A: What do you do now?
10.	T: Paste the pieces on, on the brown side.
11.	[*Later*] You make it a whole cake, like that.
12.	Ch N: Without any gaps.
13.	Ch J: OK, what do we write? On the back or on the front?
14.	T: No, underneath the quarters.
15.	I'll show you, here. See. It says, "I cut it into quarters."
16.	OK. You can copy it if you want.
17.	Ch J: I am.
18.	Ch B: This cut in half, or not?
19.	T: Yes. Cut them in half.
20.	You've already done yours.
21.	You don't need to do another one.
22.	Ch N: I've done mine, have I?
23.	T: Yes.
24.	Ch B: She's done hers. [*referring to Jane*]
25.	Ch J: [*As he writes his story*] For a big cake….

26.	Ch B: I'm going to copy it.
27.	Ch J: Will we wait till they've dried?
28.	T: No, you can bring it to me and you can tell me your story.

Here, checking interactions (lines 7, 18, 22) were concerned with folding, cutting, pasting and interpreting the shapes and parts of shapes they were making. Intentions to carry out the task accurately were supported by procedural questions (lines 5, 9, 13, 27) and clarifying statements (lines 4, 14). Child M imitated the teacher's instruction (line 3), and Child R anticipated a future instruction (line 6). Child J and Child B accepted the opportunity to copy the "story" if they wished (lines 17, 26). Evident here are further examples of children engaging in self-regulative strategies which demonstrated intentionality and integration with the knowledge base and culture of the mathematics. The children were expressing a capacity and desire for genuine co-participation and emulation of the register of the mathematics. Self-regulation was undertaken in response to others as the children inquired about and acted on correct procedures. Implicit in the conversation, too, were intentions to cooperate, and this seemed to me to be related to opportunities to show initiative and exercise a degree of choice within the constraints of an authorized set of meanings.

The Classifying Activity

After the children had completed a prescribed activity they were offered a sorting and classifying activity, involving two sets of picture cards. The teacher suggested finding matching pairs of cards, but one of the children offered another idea. She suggested a guessing game.

1.	Ch C: I'm playing a guessing game.
2.	[To T] I'll give you a clue. It's long, and it starts with "d."
3.	T: Is it alive? [She shakes her head]
4.	Ch G: I know. Dog, dog, dog, dog.
5.	T: No, it isn't alive. Do we play with it? [She shakes her head]
6.	Ch C: It's "h."
7.	T: "H." A house? Good. [To Child B, showing his work]
8.	Ch G: It can't be a house, "cause cows live in it."
9.	Ch C: "Tr."
10.	T: "Tr?"
11.	Ch G: Train?
12.	Ch C: "B."
13.	T: A boat? A bottle? What do we do with it?
14.	Ch C: We drink it.

15.   T: Water.
16.   Ch C: Wine. [*Shows the picture of a wine bottle*]
17.   T: Wine.

Here, Child C improvised a game which created opportunities for her to regulate the proceedings, beginning with a procedural statement (line 1) which encouraged other children to contribute (lines 8, 11). Child G questioned the teacher's guess of a house, within the framing of a logical explanation. Although the explanation made little sense to the teacher, it is likely it did for the child. A further challenge to conventional logic for the teacher occurred when Child C switched from thinking of something starting with "B," a bottle (line 12), to calling it wine (line 14). This did make sense, however, when the picture of a wine bottle was revealed. This experience, nevertheless, prompted the teacher to reflect on children's possible conceptual distance and what that distance might mean when adult logic is imposed on an experience. Children's processes of inductive and deductive thinking can flow and wander into quite different places to adults' logical thinking processes.

A further opportunity for improvisation occurred with another "free choice" opportunity, this time with one child playing alone with some plastic zoo animals. The teacher listened to the child's self-talk as she recorded the play interactions:

> Child L arranged animals so that they were "standing and facing each other," because one of them was going to a dance. There was a "bus" made from various shapes, and two of the animals were going to get on the bus. After the teacher had been observing the elaborate arrangement of the animals for a short time, she asked the child what she'd made. It was an animal party for the elephant "because he had his birthday, and they gave him a surprise party." Some other animals were "going to have a birthday cake," and because some others were going to the dance, they had to lie down, presumably to rest before their journey.

Within this child's representation of a real-world ritual—a birthday party—the mathematical processes of ordering and classifying were apparent, and logical, clarifying explanations were provided. The child explained what the play was about, what the animals were doing and where they were going. This observation exemplifies the power of the imagination when linked to materials that can generate meanings of purpose and significance for the participant—in this situation, these were meanings about time and place.

The Line-matching Activity

For a measuring length task the children were given a photocopied sheet consisting of four lines of different lengths and were asked to glue onto each line a strip of paper of matching length. The underlying aim of the task was to encourage use of comparative mathematical terms such as "longer," "longest," "short," and "shortest."

Some children began by anticipating how things were to be done by asking questions such as, "What're we playing?" Observing the tubs of paste, streamers and tissues one child asked, "Is it going to be a messy job?" As the children gathered the materials they needed, real-world thoughts about comparative terms, logic and precision connected one child's past experiences with the task: "You can get a medium chicken or a big chicken or a little chicken." Then, as they carried out the task, the use of mathematical terminology and demonstrations of reflective awareness were evident in the children's checking reports and inquiries:

Now I need a *real long* bit. *Not a little bit.*
I've got the *right size for the long bit.*
Look at this. Mine's *too wide.*
Now this is just the *right long* for this.
There, that's *just about the right size.*
Is this *long enough* for this bit?
Now I need *a medium bit.*

The children also argued about where to place the strips to match the given lines: "That wouldn't go over that," and "Not that one, it's too long." Later, when children completed this activity, they were given a much longer strip of paper and asked to find, outside the classroom, things that were the same length. Two children were measuring their strips of paper against the lines marked in the concrete paving when there was a discussion about whether the lines were put there deliberately, or whether they were really "cracks":

1. Ch N: It's too long, look. This one's too long. It's a long one.
2. T: Yes. They're not all the same. What will we call them, K? Cracks?
3. Ch K: Yes.
4. T: Lines?
5. Ch K: Crack.
6. T: Cracks.
7. Ch N: Cracks are from the earthquake.
8. T: That's a crack. What are those? Lines?
9. Ch N: No it isn't. That came from the earthquake.

10.   T: Really?

11.   Ch N: Yes.

Here the teacher modeled the need for precision regarding the strip of paper's length matching the lines on the concrete (which were the same length) or one of the cracks (which were of varying lengths, and mostly longer than the lines) (lines 2, 4, 6, 8). Communicating acceptance and acknowledgment of Child N's perceptual readings of the experience, and providing a sense that the meaning making process was negotiable, are evident here in the teacher's response (line 10).

SYNTHESIS AND DISCUSSION ABOUT THE MAIN IDEAS

Children Talking about What is Already Known in Relation to What is New

The children responded opportunistically to the flexibility of the socio-regulative relations in order to bring their known conceptions to the school mathematics contexts. With logic progressing only very gradually away from unreliability and unconventionality toward objectivity and validity, those types of texts that relied heavily on logic were less operational or conventional, and therefore less productive, than others. Awareness of what might be "a good reason" for believing something was only just beginning to emerge for most children (Alrø & Skovsmose, 1996).

The most prevalent genres of the children's self-initiated discourses were clarifying and procedural explanations, and reports. These would seem to be precursors to the genres involving consistent rationality, order and coherence (recounts, arguments, and evaluations). The children, as fledgling students, checked the validity or appropriateness of their own conceptions, and when they were questioned about the underlying logic of their mathematical procedures or statements, they did not, in the main, appear to be over-challenged. They could draw on strategies to conceal what they did not want to reveal. For example, they could wait for another child to answer for them, or wait for the teacher's attention to be directed toward another child or event. The use of clarifying explanations, either in isolation or in ongoing reports, was instrumental in providing clear, accessible and mostly pertinent mathematical knowledge. It was also instrumental to the generation of relevant meanings concerning awareness of the importance of listening and observing attentively, and expressing with precision and accuracy the specific details of whatever was relevant or interesting. The children were gradually becoming aware of, and using, numeracy concepts and procedures with ever-increasing sophistication.

When explanations were expressed and clarified spontaneously between students, there seemed to me to be a tendency for the children to be more conscious of the cultural agenda of the peer group. When the agenda of the teacher were less prominent, these expressions were likely to be more self-regulative than when teacher talk dominated. Tasks that were moderately challenging, such as the family graph activity, allowed the students to be more self-regulative, and provided opportunities for them to bring familiar knowledge to the contexts in which they were involved. In other words, the children's spontaneous talk provided a forum from within which their meanings became transparent and visible to themselves, their peers, and to the teacher. Furthermore, the authenticity of those meanings was more assured then than with teacher-directed discourses—that is, playing the game of saying what they thought the teacher wanted to know was inconsequential to their purposes of redescribing and reconstructing their own meanings. At the same time as they were familiarizing themselves with the formal mathematics culture, the children were attempting to locate themselves in the peer-student culture through participation in its discourses. They were shaping their peer-student identities and seeking a "fit" for their peer identity in the new student culture.

Similarly, procedural statements, questions and explanations were a rich source of information from which new meanings and ways of saying and doing things could develop and become established. Through these discourses the children were learning opportunistically. They could engage in learner demonstrations of developing procedural knowledge by employing the kinds of modeling strategies which had served them well in informal learning contexts (Lave & Wenger, 1991; Macmillan, 1999; Rowe, 1994). These processes provided opportunities to rehearse and practice the skills, strategies and procedures of the new mathematics culture, allowing students to have a sense that the learning was situated in their own activities and discourses (Lave & Wenger, 1991).

Logical and sequential systematicity was required for recounts, arguments and evaluations. In the family graph activity, for example, one child told a story about visiting her father at the weekend in the process of deciding whether she wanted to put a picture of him in her family graph. The other children helped to undo the struggle she was having about the ambiguity of her relationship with him, given that he had separated from the family. This activity stimulated considerable reflection among the group as they produced evidence for making decisions about various family members' identities and offered explanations such as: "And my poppy isn't like that or doesn't have a chair like that"; "And I do have an uncle. That might be my uncle." As yet, however, since many of the children's judgments depended on such subjective information, their capacities for objective critical reflection were still very much in a state of ontogenesis. The awareness required for these genres tended to develop from the discursive modeling and ontogenesis of "knowledgeable other" peers, as well as from the authority figures. Particular children, such as Child N, Child J, and Child A, acted as surrogate caretakers of the new mathematics register, as they challenged and assisted their peers in ways which replicated those of the teacher.

The reasoning processes involved in the move from unconscious meaning making to conscious intentionality and systematicity, were reflected in the increasing instances of deductive and inductive reasoning (Rowe, 1994). In the guessing game linked to the classifying activity, for example, the children were engaging in reasoning processes. Though its salience sometimes escaped the teacher, the engagement was offered as a genuine attempt to participate and interact with others. The students were beginning to adopt some of the new ways of perceiving experience by exercising discriminatory and anticipatory strategies, by closely monitoring their own and each other's procedures, and by observing and tuning into all the available sources of information. With intentionality, systematicity, and motivation for conscious awareness located in the social world (Bauersfeld, 1992; Lerman, 1996)—evident in the guessing game, and the zoo animal birthday party, for example—children needed opportunities to repeat actions, alter known forms, and combine known forms (Clay, 1973) in order to make new meanings. The fledgling students' modeling, imitating, and improvising discourses seemed to me to indicate that the meanings had been accessible and available to them. Opportunities for children to improvise, imagine and bring the known to the new, are most powerfully realized in self-initiated experiences. Not all of the children demonstrated the confidence or competence to do this, and nor did the activity design often permit it. After six months of being with the children for these mathematics activities, it is my contention that they were beginning to know that there were particular ways of saying and doing things. However, they did know that, while there were authorized meanings and knowledge being offered and valued, their meanings were accepted and valued and could, on the whole, be blended with the authorized meanings to create "mathematical activity" (Bishop, 1988).

Individuality, Creativity and a Sense of Agency

To enable children's natural propensities for spontaneity, curiosity, individuality, and creativity to be operational, the socioregulative relations of adults needed to be accepting and understanding of idiosyncratic interpretations of experiences (Rowe, 1994). Complicit in relations of equity and accessibility are certain understandings about the nature of the authorized meanings of the practice. In this study when the teacher accepted the students' limitations and perspectives, students tended to respond by accepting her perspectives, and whatever they perceived to be her limitations. When the teacher did not expect students to accept her ways of saying and doing things as absolutist (Alrø & Skovsmose, 1996, 1998), then they were more likely to respond to her recognitions of, and respect for, their innermost needs and desires, and accommodate and adapt to her needs and desires. Despite a desire by the researcher/teacher to accept and understand children's idiosyncratic interpretations, it was challenging to know how to protect the conventions and authorized meanings of mathematics at the same time. The process of constant review and evaluation of researcher/teacher talk contributed to the development of mediator role, more than of evaluator. A more explicit analysis of techniques to mediate has been reported elsewhere (Macmillan, 1998c, 2001).

In most conversations, children spontaneously negotiated and constructed meanings between themselves. Reporting and explaining talk typical of spontaneous interaction permitted engagement in the unconscious learning mechanisms of modeling and checking. In the line-matching activity, for example, the children articulated precise accounts of the lengths and widths of the lines as they matched the streamers to them: "a real long bit"; "the right size for the long bit," and so on. Responsiveness became a discursive practice when students modeled the socioregulative processes required to gain access to meanings and resources, and to seek, receive and give assistance. Knowledge of the operational codes of a semiotic system was supported by knowledge about positioning in relations to others in the community, and knowledge of how to "mobilize resources within the same social representation of a construct" (Lloyd & Duveen, 1993, p. 182). In the fraction activity, for example, mobilizing of the technological resources was evident for folding, cutting and pasting. In order to check that they were fulfilling the requirements of the task, the children called upon the human resources of teacher and peers. For the line matching activity the children anticipated and described the tools and procedures they would need to carry out the task. The information they needed was situated in the social dimensions of the context (Bishop, 2002; FitzSimons, 2002). The manipulative tools and perceptual materials facilitated verbal and nonverbal interactions indicating that the experiences were most accessible and salient for children when tasks stimulated perceptual interest, creativity, and were embedded in semiotic purposes and mathematical conventions.

Accuracy, Precision and the Conventions of Mathematics

Becoming literate or numerate is about developing a consciousness of abstract meanings and realizing their symbolic potential (Rowe, 1994). When an ability to perform developed in close relation to the development of a new understanding, learner demonstrations of understanding were associated with feelings of success (Rowe, 1994; Lave & Wenger, 1991).

Given that these children were only just entering a formal mathematics education program that offered opportunities to tune into meanings about the necessity for accuracy and precision, their open, fluid and spontaneous interactions seemed to indicate confidence that their offerings allowed them to identify fully with the new culture's codes and meanings. Success is more explicitly evident in the fraction activity, for example, when the child says, "I done it. I'm having a brown handkerchief."

Choice and challenge—opportunities for intrinsic motivations to be sustained (Malone & Lepper, 1987)—were operational within contexts in which flexibility and space to discover and explore new meanings were provided. In the family graph activity, for example, the children could converse freely with each other and negotiate sensible choices. Interactions such as, "That might be my uncle" was followed up by another child, "Yes, this might be my uncle too. 'Cause my uncle has a moustache." As the children discovered similarities and differences between their own family members and the pictures given to them, the young students were directly confronted with the symbolic potential of mathematics—its numerical, spatial and measurement systems.

Knowledge of the counting systems and awareness of the importance of print was revealed incidentally and spontaneously by some of the children. That knowledge and awareness were expressed in reading, writing and speaking. Use of precise identifying attributes as descriptors of phenomena in spontaneous discourses indicated growing understanding of their significance: for example, "a little bit," "only one more left," "that's a medium bit," "is that the same size," "a bit little," and "real little." Use of the correct terminology to label and classify objects was also being adopted (e.g., the lines or cracks in the concrete).

Negotiability of Meanings and Procedures

Opportunities for the students to negotiate their own ways through meaning making and knowledge of mathematical procedures were associated with access to socioregulative mechanisms. The major factors affecting possibilities for children to find their own ways through tasks at school are task design and small group or whole class organization. The two experiences reported here that offered the highest flexibility for the children and the least need for supervision and control by the teacher were the guessing game and the zoo animal play. These were open-ended, with only the materials being chosen by the teacher. The children could determine the process and outcome. For other experiences, the mathematical content and conception of a product were devised and planned by the class teacher. The processes assumed degrees of negotiability mostly because there were only four or five children in the group and one teacher stayed with them throughout, providing them with close and undivided attention. It seems to me that the small group arrangement allowed the children to exercise responsible self-regulating practices and strategies—verbally, at least—in ways that may not have been possible with one teacher and thirty children. Apart from the teacher's usually brief introductory segment for each group, the children could speak openly and freely, and they could observe closely and listen to others talking. They felt free to inquire about each other's and the teacher's conceptual and procedural knowledge at any time they wished. Compared to a whole-class group, the small group also afforded a degree of privacy for the correction of procedures. With waiting time shorter and turn-taking less important, the interactional context could be relatively informal. In essence, then, flexible socioregulative relations were made possible by the small-group structure of the lesson sessions.

The small group organization allowed the teacher to attend to, observe, and monitor the children closely. Paradoxically, though, it also allowed for concealment of uncertainties or shortfalls in understanding, as the children could respond on each other's behalf as they had access to each other's procedures. Bauersfeld (1992) emphasized how useful it can be for students to read the written solutions of others so that they can develop the kinds of reflection strategies relevant to their own work. Such learning by "contrasting" can lead to meaning being based on "the functioning of a network of personal actions and descriptions rather than on single words or symbols" (Bauersfeld, 1992, p. 469).

Support of each other through explaining, discussing and reflecting allowed the challenges of the mathematics to be alleviated. Socially generated meanings left "enough space for individuals to find their own path through the "data" and for intelligent curtailments" (Bauersfeld, 1992, p. 476). A focus on processes at least as much as on outcome within a context of authorized meanings allowed children to maintain their urges for engagement, involvement and control (Bauersfeld, 1992). As "expert" emulators of mathematics authority figures, particular children, such as Child N, Child, A, Child J, and Child M, acted as models of numerate awareness, understanding and competence. The small group context provided opportunities for them to be powerful models to other students—to bridge the gap between where they were in their understandings and perceptions and where the teacher was. In the fraction activity, for example, Child N elaborated on the teacher's instruction, "You make it a whole cake like that" by adding "without any gaps."

These children were able to argue and be assertive. This was evident in the family graph activity, for example, when a series of interchanges about identifying attributes of family members, and comments such as "Your brother doesn't look like that" was followed by, "I know, but he has a hat like this but a kind of different color." A combination of confidence and understanding gave them a clear passage into the meanings and relations of the practice. Their peers were likely to want to emulate these children, especially when they were accepted or protected by them. The context allowed these knowledgeable children *to be*, rather than having *to be taught.* These "student-teachers," or "surrogate teachers," identified with the teacher's role and adopted her perspectives because they understood and could accept the meanings and relations being communicated directly and indirectly by her.

In relation to the role of the teacher as monitor and mediator of peer negotiations, there were opportunities for her to take a "back seat" or non-participant position, providing support and guidance only when requested by the children. The students, in turn, within her gaze and guidance, negotiated their own ways through new meanings and procedures.

MODELING AND PARTICIPATING IN DISCOURSES AND ACTIVITIES

In this classroom, the teacher was modeling how to make meaning as she sought and provided clarifying, procedural, and logical explanations, statements and questions. At the same time, she modeled how the participants of the community of learners could and should relate to each other as social beings. For the emerging student identities, social and mathematical meanings were intermingled with each other, and meaning making was situated in active, involved participation (Lave & Wenger, 1991; Lerman, 1996; Nunes, 1996).

Teacher and students echoed each other's explanations in an elaborate and intricate system of language-patterned modeling that opened the paths toward literate meanings. The meanings to be discovered and experienced along such paths were not conveyed primarily through direct instruction. Rather, numeracy events were woven into the "fabric of classroom life," and evolved "in the course of, and in support of the particular purposes and themes of a school experience" (Kantor, Miller, & Fernie, 1992, p. 200). This was evident in the family graph activity through the children's stories of their families, and in particular, their understandings of identifying characteristics of family members, and their representation of relationships between family members. In such "deformalized" formal contexts, there was no single path toward numeracy. These multifaceted and transformative paths linked numerate action with social action. Students became predisposed to "empower action, feeling, and thinking in the context of purposeful social activity" (Wells, 1990, p. 14).

The degrees to which individual students were able to access the meanings through involved participation in the activities and discourses depended to some extent on the familiarity of such patterns of language. For those children for whom such discourses were familiar and natural, socioregulative mechanisms could also function optimally. However, for other children emulation was less natural and unconscious. Vygotsky's (1978) suggestion that social situations were always "in advance of independent activity" (p. 57), has been taken up by Rowe (1994, 1998) in her argument that it is recognition of a difference between personal meanings and those of others which prompts change, growth and development. Children who struggled to express their understanding were gradually beginning to notice mismatches between their own and others' understandings of the intended meanings, procedures and purposes of the activities. In the line matching activity, for example, differences and similarities between the teacher's and the children's understandings of a "line" (the formal mathematical concept) and "a crack" (a concrete perceptual concept) became apparent.

CONCLUDING COMMENTS

In offering a small glimpse into one classroom's mathematical activities at the beginning of school, I have endeavored, through socio-linguistic and sociocultural lenses, to highlight understandings about children's competence with language and their use of various types of text to access knowledge about numerate purposes and the mathematical meanings underlying them. Fundamental to that competence is the cultural situation. Within that situation mathematical meanings are negotiated, even as mathematical precision and accuracy are protected.

Children are immersed in the numerate meanings of the culture (Bishop, 2002; FitzSimons, 2002) and immersed in modes of active participation, perceptual and intellectual engagement. Teacher talk blends with fluid and sustained learner talk. In this investigation, by talking and listening, and by monitoring and supporting each other, as they acted out the ways of the numerate culture, children became empowered to "take up" numerate identities. They demonstrated how they benefitted from explaining, reporting, recounting, arguing, discussing and being critically reflective. They revealed how communicative competence is integrally related to social competence, and how both of these provide access to the development of literate and numerate identities (Sierpinska, 1994; Sierpinska & Kilpatrick, 1998; Steinbring, Bartolini Bussi, & Sierpinska, 1998). What became clear in the investigation was the way in which individual intentionality and culturally-coded systematicity merge and dwell in comfortable and fruitful dynamic relations with each other.

APPENDIX

Table 5.1. Definitions of Oral Text Types

Text type	Function	Structure	Summary
Logical Explanation	To explain the processes involved in the formation or workings of natural or social phenomena.	A sequenced explanation of why or how something occurs.	Reason why a judgment has been made.
Procedural Explanation (Instruction)	To describe how something is accomplished through a sequence of actions or steps.	Expresses a goal, the materials, and the steps for achieving the goal.	How something is done.
Clarifying Explanation	To describe a particular person, place or thing.	Identifies a phenomenon and describes parts, qualities, characteristics.	What a particular thing is like, or what is needed.
Report	To describe the way things are, with reference to man-made or natural phenomena.	Tells what the phenomena are and what they are like—parts, qualities, behaviors, uses.	What an entire class of things is like.
Recount (Anecdote)	To retell an event for the purpose of informing or entertaining.	Sets the scene; tells what happened in sequence.	What happened in a particular situation.
Narrative	To retell a humorous or problematic event.	Sets the scene, evaluates the plight, explains the complication, resolves the crisis.	An interpretation of what happened to someone else.
Argument (Exposition)	To persuade that something is the case.	Introduces the subject matter, indicates speaker's position, outlines main arguments.	Arguments to do with why a thesis statement has been produced.
Discussion	To present two or more points of view about an issue.	States the issue, argues for and against, forms a conclusion or recommendation.	Arguments for and against an assumption or idea.
Evaluate	To critique an event, object, situation or experience.	Provides a contextual orientation, an interpretive recount, and an evaluation.	Personal, critical, reflective assessment of phenomena.

Table 5.2.   Definitions of Socio-cultural Modeling Strategies

Modeling Strategies		
Category	*Definition*	*Examples*
Modeling	Demonstrations of interactions or procedures that represent culturally appropriate ways of saying or doing things.	"Do we have to fill up all of these holes?" "Put your hand up if you've got orange."
Imitating	Saying or doing the same as another.	
Checking and Testing	A tentative inquiry as to the soundness of an idea or procedure.	"Are these pieces longer than those?"
Assisting	Demonstrations of help or support for another person.	"Can you do this up?" "Will you help me make this?"
Observing	Watching with the aim of understanding a concept, procedure, event or behavior.	Molly watches Trish cutting. Peter, Robert and Tim are in the hospital area, just looking.
Anticipating	Predictions of what can be said or done based on sound conceptions.	"Will we put our names on the papers?"
Positioning	Physically and metaphorically locating oneself in an advantageous position	"I'm sitting beside you."
Improvising	Improvisation involves interpretation of a basic idea and then readapting it in some way in order to extend interest and involvement. It is closely connected to Bishop's (1988) playing activity.	"Can we have some black dresses and brooms? You could get them before you come to pre-school. We want to dress up as witches."

Table 5.3. Definitions of Socio-cultural Outcomes

Socio-Cultural Outcomes

Category	Definition	Examples
Clear Access	Demonstrations of a capacity to provide clear access to the kinds of social and conceptual understanding which allow culturally acceptable behavior to develop.	"You can make a little ring with that." "Do you know what my mum's name is?"
Co-participation	Demonstrations of a capacity to contribute equally in an activity by sharing the control of procedures.	"Do you want me to put the matching out for you?"
Responsible Self-Regulation	Demonstrations of a capacity to be responsible in the use of interpersonal choices.	"It's time to pack up." "Where are we supposed to go?" "Why did you say that bad word, Paddy?"
Self-agency	Demonstrations of capacities to be self-regulative, to have choice and control over one's actions and interactions.	"I'd like to play a guessing game with these cards. I've already built something with the Unifix blocks."

REFERENCES

Alrø, H., & Skovsmose, O. (1996). On the right track. *For the Learning of Mathematics, 16*(1), 2–8.

Alrø, H., & Skovsmose, O. (1998). That was not the intention! Communication in mathematics education. *For the Learning of Mathematics, 18*(2), 42–51.

Australian Association of Mathematics Teachers. (1997). *Numeracy = everyone's business.* Adelaide: AAMT.

Australian Association of Mathematics Teachers. (1998). *Policy on numeracy education in schools.* Adelaide: AAMT.

Bauersfeld, H. (1992). Classroom cultures from a social constructivist's perspective. *Educational Studies in Mathematics, 23*, 467–481.

Bishop, A.J. (1988). *Mathematical enculturation.* Kluwer: Dordrecht.

Bishop, A.J. (2002). Critical challenges in researching cultural issues in mathematics education. *Journal of Intercultural Studies, 23*(2), 119–131.

Bourdieu, P. (1977). *Outline of a theory of practice.* London: Cambridge University Press.

Chaiklin, S., & Lave, J. (Eds.) (1993). *Understanding practice: Perspectives on activity and context.* Cambridge, CT: Cambridge University Press.

Clay, M. (1973). *Reading: The patterning of complex behaviour.* Auckland: Heinemann.

Davies, B. (1993). *Shards of glass. Children reading and writing beyond gendered identities.* Sydney: Allen & Unwin.

Davies, B., & Harré, R. (1990). Positioning: The discursive production of selves. *Journal for the Theory of Social Behaviour, 20*(10), 43–63.

FitzSimons, G. (2002). Introduction: Cultural aspects of mathematics education. *Journal of Intercultural Studies, 23*(2), 110–118.

Gerot, L., & Wignell, P. (1994). *Making sense of functional grammar: An introductory workbook.* Cammeray: Antipodean Educational Enterprises.

Halliday, M.A.K. (1994). *An introduction to functional grammar.* London: Edward Arnold.

Kantor, R., Miller, S.M., & Fernie, D. (1992). Diverse paths to literacy in a pre-school classroom: A sociocultural perspective. *Reading Research Quarterly, 27*(3), 185–201.

Lave, J., & Wenger, E. (1991). *Legitimate peripheral participation: Situated learning.* Cambridge: Cambridge University Press.

Lerman, S. (1996). Intersubjectivity in mathematics learning: A challenge to the radical constructivist paradigm? *Journal for Research in Mathematics Education, 27*(2), 133–150.

Lloyd, B., & Duveen, G. (1993). The development of social representations. In C. Pratt & A.F. Garton (Eds.), *Systems of representation in children's development and use* (pp. 167–183). Chichester: Jacaranda Wiley.

Luke, A. (1993). The social construction of literacy in the primary school. In L. Unsworth (Ed.), *Literacy, learning and teaching: Language as social practice in the primary school* (pp. 1–54). Crows Nest, NSW: Macmillan Education.

Macmillan, A. (1997). *An investigation of the initiation process of pre-school children into the culture of a formal mathematics education class*room. Unpublished PhD thesis, The University of Newcastle.

Macmillan, A. (1998a). Investigating the mathematical thinking of young children: Some methodological and theoretical issues. In A. McIntosh & N. Ellerton (Eds.), *Research in mathematics education: A contemporary perspective* (pp. 108–133). Perth: MASTEC.

Macmillan, A. (1998b). Pre-school children's informal mathematical discourses. *Early Child Development and Care, 140,* 53–71.

Macmillan, A. (1998c). Responsiveness as a viable interpersonal relation in teacher-directed discourses. *Early Child Development and Care, 140,* 95–113.

Macmillan, A. (1999). Pre-school children as mathematical meaning makers. *Australian Journal for Research in Early Childhood, 2*(6), 177–191.

Macmillan, A. (2001). *Deconstructing social and cultural meanings. A model for educational research using postmodern constructs.* Melbourne: Common Ground Publishing.

Malone, T.W., & Lepper, M. (1987). Making learning fun. In R.E. Snow & M.J. Farr (Eds.), *Aptitude, learning and instruction, Vol. 3: Conative and affective process analysis* (pp. 223–253). London: Lawrence Erlbaum.

Marks, G., & Mousley, J. (1990). Mathematics education and genre: Dare we make the process writing mistake again? *Language and Education, 4*(2), 117–135.

Noss, R. (1998). New numeracies for a technological culture. *For the Learning of Mathematics, 18*(2), 2–12.

Nunes, T. (1996). Language and the socialization of thinking. In L. Puig & A. Gutiérrez (Eds.), *Proceedings of the twentieth annual meeting of the International Group for the Psychology of Mathematics Education* (Vol 1, pp. 71–77). Valencia, Spain: Dept. de Didàctica de la Matemàetica, Universitat de València.

Peirce, C.S. (1955). Logic as a semiotic: A theory of signs. In J. Buchler (Ed.), *Philosophical writings of Peirce* (pp. 98–119). New York: Dover.

Peirce, C.S. (1966). *Collected papers of Charles Sanders Peirce.* Cambridge, MA: Harvard University Press.

Resnick, L.B., Levine, J.M., & Teasley, S.D. (Eds.) (1991). *Perspectives on socially shared cognition.* Washington, DC: American Psychological Society.

Rowe, D.W. (1994). *Preschoolers as authors: Literacy learning in the social world of the classroom.* Cresskill, NJ: Hampton Press.

Rowe, D.W. (1998). Examining teacher talk: Revealing hidden boundaries for curricular change. *Language Arts, 75*(2), 103–107.

Sierpinska, A. (1994). *Understanding in mathematics.* London: The Falmer Press.

Sierpinska, A., & Kilpatrick, J. (1998). *Mathematics education as a research domain: A search for identity. An ICMI Study.* Dordrecht: Kluwer.

Steinbring, H., Bartolini Bussi, M., Sierpinska, A. (1998). *Language and communication in the mathematics classroom.* Reston, VA: Council of Teachers of Mathematics.

Vygotsky, L. (1978). *Mind in society: The development of higher psychological processes.* Cambridge, MA: Harvard University Press.

Wells, G. (1990). Creating conditions to encourage literate thinking. *Educational Leadership, 47,* 13–17.

CHAPTER 6

"THERE'S NO HIDING PLACE"

Foucault's Notion of Normalization at Work in a Mathematics Lesson

Tansy Hardy

ABSTRACT

In this chapter I look at the appropriateness of theoretical frameworks used in mathematics education research. I outline an analytical strategy drawn from Foucault's analytics of power as a productive way of investigating the object of our study. I show how this strategy can be used to reflect on the effects of mathematics education discourse. In particular I discuss a video extract for teachers and suggest how it might uncover hidden facets of classroom practices and how it can assist in identifying possible effects of these practices on learners and teachers.

INTRODUCTION

In this chapter I discuss my consideration of the appropriateness of theoretical frameworks used in mathematics education research. I argue that as researchers, not only do we need to confront the inevitable cultural nature

Mathematics Education Within the Postmodern, pages 103–119
Copyright © 2004 by Information Age Publishing
All rights of reproduction in any form reserved.

of our subject of study, but we also need to work with the provisionality of our knowledge and understanding in this area.

I outline an analytical strategy that acknowledges the complex cultural nature of mathematics education practices—a "toolkit" drawn from Foucault's analysis of power. I show how this can be used to reflect on the effects of mathematics classroom discourse, in particular, an extract from video classroom exemplars. The video exemplars were produced as part of a professional development package to guide English primary school teachers on key aspects of desired "effective mathematics teaching" in the most recent United Kingdom (UK) mathematics school curriculum "innovation"—the National Numeracy Strategy (DfEE, 1999). In relation to this highly monitored project, I discuss how Foucault's analytical strategy might be employed to uncover previously camouflaged and hard-to-pin-down facets of mathematics classroom practices, and how it might also assist in identifying possible effects of these practices on learners and teachers.

MY INTENTIONS

It has been noted that little significant change has been achieved in UK mathematics education practices despite such major reform initiatives in the 1980s and 1990s of the National Curriculum and the General Certificate of Secondary Education (GCSE) examination. In particular, there has been a persistent failure to help particular groups of learners learn school mathematics (Askew, 1996; Brown, Askew, Baker, Denvir, & Millett, 1998). This is paralleled in my experiences working in initial and in-service teacher development since 1983 (Hardy, 1996). The recent implementation of the "National Numeracy Strategy" throughout England, has resulted in a recognizable change in the structure of lessons in primary schools (for children aged 5–11) and to some extent in a change in how teachers talk about their teaching in mathematics (Earl, Levin, Leithwood, Fullan, & Watson, 2001). Further, other researchers from the UK, Australia, and the USA have commented that despite cosmetic changes, students persistently experience a fundamental curriculum of fact learning and routinized computations where they are expected to be consumers of established mathematical "truths" (e.g., Klein, 1998, 1999).

It is pertinent then to look carefully at the nature of this English based large scale reform and ask what kinds of effects will be produced in the learners of mathematics and their relationship to this school subject. Who will gain and who will lose in this new arrangement?

Jones asked this question as a teacher researcher in a primary school in inner city Manchester. She had identified the inescapable failure produced for certain groups of children by their experience of learning mathematics

(Jones & Brown, 1999). Her research focused primarily on how the condition of girls is constructed in nursery level education and she argued that even in nursery level education, understandings of gender roles were substantially developed and that many play activities served to reinforce these. She explored ways in which alternative teaching strategies might assist in eroding existing norms. However she found herself caught in a personal debate about whether it was better to engage the children in a critical education program in which some of these oppressive norms were challenged, or whether she should embrace the "back to basics" campaign that was then promulgated, to ensure that the children achieved threshold levels of achievement. Her conclusion was that the children concerned would end up on the bottom of the heap, whatever she did. She found herself moving from an emancipatory attitude in which she sought to improve the children's lot to what might be said to be a more postmodernist fatalism.

To work on such concerns, I have argued that those of us undertaking mathematics education research should think hard about the nature of the objects of our studies, our concepts and our methods of study (Hardy & Cotton, 2000). Not least we should recognize that our work is profoundly cultural and political. This prompts the search for analytical strategies that might reveal the means through which this repeated failure to learn mathematics is brought about. Dunne (1999) provides another example of this search. She finds a paucity of research that adequately considers social or cultural aspects in mathematics education literature and asserts the inadequacy of applications of some theories of learning in revealing the cultural nature of mathematics education.

> Within recent work, including social constructivism, the notion of classroom culture carefully circumscribes its concern as within the confines of the classroom, sometimes as if disconnected from any external influences, for example, what the students bring with them into the classroom. This construction of "culture" is evidently too limiting for the development of mathematics education research with a social justice concern. Such work necessarily connects the micro- to the macro-level; local practices to policy; individuals to communities. (Dunne, 1999, p. 117)

My starting point for such concerns is to identify ways of becoming aware of such effects for my practices as mathematics teacher and researcher. This is also a central concern for Meaney (this volume) in her investigation of a curriculum development project. Becoming aware of the effect of one's practices is not a straightforward task. Foucault pointed this out when he said: "People know what they do; they frequently know why they do what they do; but what they don't know is what they do does" (Foucault & Deleuze, 1972, p. 208).

For example, it has been suggested (Bourdieu & Passeron, 1977) that schools serve to reproduce the existing injustices in society through practices seen as common sense in school, but which are based on the class structure present in society. It is this trick of power to masquerade as "common sense" that leaves us unaware of the effects of our practices. For my own work this has led to an exploration of the form of the cultural practices that make up school mathematics education and a search for theoretical frames that will lead to accounts of the effects of these practices, for how these practices gain their power. Work that acknowledges the cultural-political nature of teaching and learning and that offers insights into how our practices of school mathematics and their effects come about is becoming more prevalent in mathematics education research (e.g., Appelbaum, 1995; Klein, 1998; Lerman, 1998; Valero & Vithal, 1998; Walkerdine, 1994; Zevenbergen, 1996). This follows the hope of Ball (Ball, 1990, cited in Roth, 1992) that putting Foucault to good use in education would unmask the politics that underlie some of the apparent neutrality of education reform.

MY THEORETICAL FRAMEWORK: DISCURSIVE PRACTICES

I have found that frameworks, where the actors, objects and concepts of study are seen to have been constructed through "discourse," are particularly helpful in recognizing the complex nature of mathematics education practices and in revealing "taken as natural" aspects of those practices. I take the term "discourse" in its broadest sense, referring to anything communicated using signs (including actions and interactions in the classroom, resources used, and arrangements of the furniture). I will work with Foucault's development of this to examine what he calls "discursive practices" in society. The modifier "discursive" stresses the ways in which all practices are bound up in systems of knowledge. The playing out of these "knowledge producing" systems, which infuse everyday activities, shape the experience of being human.

Foucault argues that these discursive practices differentiate people in relation to cultural norms that have become regulatory and often self-regulatory ways of knowing (Foucault, 1972). Foregrounding the discursive nature of mathematics education practices marks a shift away from viewing culturally accepted norms as perpetuated through traditions and sees knowledge as produced through a process of describing and ordering things in particular ways. It is this process that produces "subjects." "Subjects' here are understood in both senses, as persons and bodies of knowledge.

For example, the categorization embedded in the 1995 constitution of The National Curriculum for Mathematics in the UK (DfE, 1995)—

ordered into eight levels and four attainment targets—became a cultural norm, regulated through descriptions. For several years after the introduction of this curriculum document, these descriptions came to be taken as natural and obvious ways of categorizing mathematics. These persisted until the major reformatting of the mathematics curriculum when the National Numeracy Strategy in 1999 (DfEE, 1999) was imposed.

The theoretical notions of discursive practices and the producing of a subject enable us to become aware of the constitutive nature of mathematics education classifications and descriptions, and to think about how things might be described differently.

POWER IS PRODUCTIVE

When I strive to consider the effects that school mathematics discursive practices have on both learners and teachers, Foucault's (1972) discussion of the productive nature of modern power proves valuable. He did not treat power as a commodity that can be owned or exchanged; that is, it is not invested in one individual to exert influence over another. Rather he directed us to its exercise, interpreting power as dispersed, dynamic and productive, and to its classificatory procedures that objectify and constitute "subjects." This process is described by Pignatelli (1993) in his work on teacher agency: "The obedient subject is produced and sustained by a power faintly noticed and difficult to expose; a power that circulates through these small techniques among a network of social institutions such as the school" (p. 420).

In Foucault's words:

> If power were never anything but repressive, if it never did anything but to say no, do you really think one would be brought to obey it? What makes power hold good, what makes it accepted, is simply the fact that it doesn't only weigh on us as a force that says no, but that it traverses and produces things, it induces pleasures, forms, knowledge; it produces discourse. It needs to be considered as a productive network, which runs through the whole social body, much more than as a negative instance whose function is repressive. (Foucault in Rabinow, 1986 p. 61)

Roth (1992) promotes the importance of the positive, enabling voice of Foucault's power: "Insofar as power specifies boundaries of pleasure as well as ranges of normality, it is not wholly repressive. In fact, power requires that resistance to power is always possible" (p. 690). It requires "that 'the other' (the one over whom power is exercised) be thoroughly recognized and maintained to the very end as a person who acts" (Foucault in Dreyfus and Rabinow, 1982, p. 220). Roth criticizes those who dwell solely on limiting aspects of the exercising of power. He claims that working with both sides of the contradictory nature of power is key to understanding the effects and persistent acceptance of pedagogic practices of educational institutions.

From this it can be seen that an analysis taking this approach can attend to the discursive nature of professional practices—to how simultaneously human beings (teachers and children) are defined by discourse's use, while at the same time the discourse describes them. In this perspective, organized forms of knowledge, working together with their associated institutions, have significant effects on people and their possible actions. These effects can be both repressing and enabling.

I have experienced the "defining effect" of discourse when, as a researcher observing in a classroom, a child tried to elicit "my help." For that child a characteristic of a "teacher" might have been "one who helps." I would define myself as "teacher," or not, by my response to her. At the same time, through that response, I would contribute to the definition of what a teacher is and does. I have entered the room with labels "nonparticipant observer" and "researcher" in mind. I know, however, that I had not shed my "teacher" self as I passed through the door. I became acutely aware that my response would define who I was and what I was in this classroom context. It would also determine my possible actions in that arena.

Further to this, the most effective form of regulation is self-regulation. This refers to our willingness to accept the determination and limitations of what we can know about ourselves and how we might act, as natural or inevitable conditions.

> Agencies of power sustain control not by repression but by eliciting consent. People in modern institutions are conditioned to accept being an object to others and a subject to themselves. The very processes we use to inscribe our self to our self put us at the disposition of others. The task of creating rational, autonomous persons falls initially to pedagogical institutions; their goal is to produce young bodies and minds that are self-governing; failing that, they try to make their graduates governable. (Roth, 1992, p. 691)

Through a Foucauldian analysis of mathematics education discursive practices, I can look for traces of the productive and determining effects of power; at what enables actors to act and to take a particular position and so to be heard. I can also "sniff out" and challenge the conditions that present themselves as inevitable and incontestable.

MY EXPLORATION OF AN ANALYTICAL STRATEGY

The next section is offered as an example and as a trial of an analytical strategy drawn from the theoretical discussion above. For this particular strategy, I outline a "toolkit" drawn from Foucault's analysis of power and the production of knowledge, and use this to examine mathematics classroom discourse. An extract from the video materials that were used as part of training days for all English primary school teachers as part of the introduction of the National Numeracy Strategy (DfEE, 1999) is used as a short exemplar for such an analysis. This video material was also published as a part of the "pilot" National Numeracy Project (NNP, 1998). The National Numeracy Strategy (DfEE, 1999), the latest UK major initiative, can be seen as foregrounding the teacher's role in the child's learning of mathematics. Brown, Jones and Bibby (this volume) have explored what implications the revised teacher's role has upon trainee teachers' and beginning teachers' work. The reforms provide a timely context for my investigation. I examine the effects of this new emphasis upon the actors (here teachers and learners) in the mathematics education arena.

MY "TOOLKIT": TECHNIQUES OF POWER AND NORMALIZATION

Any mathematics education discourse positions and categorizes children and teachers in particular ways (Hardy, 2000). For example, children are often portrayed as the ones who "have difficulties" or "misconceptions." This pathologizing can be brought about through particular techniques of power linked to the process of "normalization."

Normalization is the mechanism that categorizes people into normal and abnormal. Connecting the notion of normalization with power reveals the process that determines what becomes valid knowledge in the classroom and how that knowledge can be expressed and by whom. It is the normalization process that determines, for example, who "has the difficulties" and who does not. Foucault (1977) also claims that examination "… is a normalizing gaze, a surveillance that makes it possible to qualify, to classify, and to punish. It establishes over individuals a visibility through which one differentiates them and judges them" (p. 184).

Specific forms of surveillance bring about normalization, such as individualization and totalization. For totalization, a group specification is given, asserting a collective character. This is a readily recognizable element of pedagogic activity where "we" or a class name is used to address whole groups of participants. For example, a teacher may praise a whole class by saying: "Well done, 3W. I'm pleased with the way that you moved back to your desks." Individuals and their behavior are ignored by this statement. It regulates the group behavior and asserts group characteristics. A child claiming an individual voice could find herself excluded from the group and from the classroom culture. Individualization is the technique of giving individual character to oneself and may be an attempt to resist unwelcome totalization. However it can also be a way of drawing attention to a child's deviance from the classroom norms and establishing their abnormality—again, a common classroom practice.

Foucault invites us to take these particular "techniques of power" to form an analytical "toolkit" to look at how power relations function at the micro-level and to reveal the constructed nature of institutional practices (Foucault, 1977, p. 218). Foucault uses the words "techniques" and "technology" when referring to the physical or mental act of constructing reality. This reemphasizes the way that human beings are essentially constructed by the seemingly non-discursive background practices into which they are thrown.

I will treat the video extract given below as a text that has been produced within a particular discursive practice—an exemplar of teaching practices promoted by the National Numeracy Strategy team—that can be analyzed at the micro-level in the search for patterns of practice.

My toolkit is formed from the Foucauldian notions of "power as productive" and "normalization"; in particular the notion of "surveillance as a normalizing gaze." I aim to demonstrate that this toolkit can be used to identify facets of the classroom practices of this "numeracy lesson" that would otherwise be easily overlooked. This will lead to a particular "reading" of the classroom interactions and teacher's descriptions of her work, a reading that may seldom have been voiced in mathematics education literature.

TRANSCRIPT OF VIDEO EXTRACT

This extract (DfEE, 1999; NNP, 1998) is made up of interview scenes where the teacher comments on her classroom practices. These are interspersed with classroom scenes where the teacher moves around the room asking short calculation questions of the whole class. Her questions and instructions are presented to the whole class, whether she is referring to the children as a whole group or as individuals. The children's desks are arranged in blocks of six and each child has two sets of cards, both numbered from 0 to 9, in front of them. They hold up cards to show their answers to the questions asked of them.

Comment from Teacher:
A few children don't put their hands up. They try to hide, but that's the

idea. There is no hiding place. You encourage them all as long as you give them positive feedback. Even if they get it wrong, they are not scared to give an answer.

In Classroom Scene:

Teacher: "Show me a multiple of five bigger than 75 ... Is that a multiple of five though, Michael? It's bigger than 75 but check it's a multiple of five...

Well done Sarah!

Teacher: Show me three threes ...

Three threes? Check again please, Lauren.

Check please, Joe. You are looking at someone else's. Don't just look at someone else's. If you're not sure get your fingers and count in lots of three. Let's do it together (chanting) three, six, nine. You should be showing me nine there.

Comment from Teacher:

Some children don't have instant recall of three threes but I've given them a method to work it out. "Get your fingers and count in threes." So as long as they do regular counting in threes and they've got that pattern, they have got a method to do it. When I see the children struggling I take them back to the method or strategy that we've talked through together to help them through that. They are not stood in queues waiting to get a book marked; they are getting instant feedback. They are not scared to get an answer wrong. They're having a go, they are risking things, and you don't gain anything unless you have a few risks and that's what they are doing.

In Classroom Scene:

Teacher: Have a quick check of that one, Misha. You should be showing me twelve.

Comment from Teacher:

It really works. We've seen it work. The children are motivated. The children want to learn. You never have to tell children "Are you messing around?" They're not. They are trying. They might not be succeeding but they are trying. They really love the pace. Children don't like sitting for 20, 30 minutes on one task especially if they are struggling on it. This doesn't allow that. The children have to find answers. They work together. They help each other but they are also pushing forward. The task is changing all the time. As long as you stay focused on target, most lessons you achieve eighty percent of children come out learning something that they didn't go in knowing and that's a wonderful experience and encourages you to go on further.

A FOUCAULDIAN READING

Surveillance: The normalizing gaze. This is brought about through the bodily position of each child. By holding their cards up, by putting their hands up, they expose themselves to this surveillance. "There is no hiding place." By holding up their answer on the cards or by offering their answer verbally they identify themselves with their answers. They are judged as individuals by the rightness or wrongness of their answers. It is the child who is wrong not just the answer. This "wrongness" becomes part of whom they are, and this aspect of their subjectivity is produced.

"Have a quick check of that one, Misha. You should be showing me twelve." This jars somewhat with the teacher's comments on encouraging them all. She states that by giving them positive feedback they are not scared to give an answer even if they are wrong. "They're having a go, they are risking things and you don't gain anything unless you have a few risks and that's what they are doing." In order to be included in this classroom culture the children have to be prepared to reveal themselves, to take risks, to risk being declared wrong/not normal and so excluded. It is worth asking who runs the biggest risk here. For example, in the classroom scenes on the video it can be seen that significantly more boys than girls are raising their cards and that they do so more rapidly.

The teacher believes that she surveys them all—"There is no hiding place." However it may be that some children escape her gaze, either because of the strength of her assumption that they are all willing to have a go or because it is not practicable for the teacher to look everywhere at all times. It may be that particular children can use this to "resist" running the risk of being "wrong." They could choose to respond slowly, make motions to consider which of their cards to display but avoid a final selection. Certainly, it is not the teacher who has to take risks with right and wrong answers here. She is not seen here to follow her own adage: "you don't gain anything unless you have a few risks."

Of course overlapping and colliding discourses are involved in the production of the subjectivities of both the learners and the teacher. As subjects they acquire an assortment of overlapping ranks and labels as they come to know what they are (knowledge produced through the technique of surveillance). These can differentiate them in multiple ways. The subject is not fixed for all times; it is a flexible and segmented presence.

This can be seen for the teacher. There is a strand of "caring" running through the teacher's descriptions: "Let's do it together." You are not on your own. "You should be showing me 9"—a safe process: predict and see if you are right. There is an emphasis on how making things familiar helps them cope with the risk taking. "[T]hey do regular counting in threes and they've got that pattern ... I take them back to the method or strategy that we've talked through together to help them through that. She promotes "instant feedback." This would make practices familiar more quickly so that their fear of getting an answer wrong is minimized. She watches and listens to the children, both evaluating the children and monitoring the risk level. She recognizes the debilitating effect of "too much risk"—"if they are struggling"—and describes strategies she can use to help them feel better.

But I can see a simultaneous insistence that the children expose themselves to risk that conflicts with her care for their mental health. Is the teacher offering comfort by holding the child's shaking hand as she pushes her to the boundary (brink)? Does this "care" make the normalizing process and the risk of being wrong and marginalized more justified? Acceptable? More bearable? "The more "humane" we make these institutions, the more insidious has been the exercise of power" (Roth, 1992, p. 691).

Individualization and totalization. The teacher talks of "positive feedback." I can look at the classroom practices and ask what feedback is given and what constitutes positive. There are appeals to pace, quick responses, and instant feedback in terms of right and wrong, *all* are given feedback, no waiting, and no struggling for a sustained length of time. Individual children are told when they are wrong, albeit with some gentleness. The teacher might claim that she uses children's names to give praise and to encourage them to try again "Three threes? Check again, please, Lauren." The effect of this individualization may be different. Is it positive feedback or is it obvious to Lauren that her answer is wrong? Does "check again" soften the effect of being labeled abnormal and forced to the margins?

This is more marked for Joe, "Check please, Joe. You are looking at someone else's. Don't just look at someone else's." The knowledge constructed through these discursive practices is not just about right answers but there is a clear production of "wrong behavior" and a privileging of one particular "right method": "If you're not sure get your fingers and count in lots of three. Let's do it together (chanting) three, six, nine. You should be showing me nine there." In my reading of the situation, the totalization of "doing it together" will help define this as *the* valid method to use if a child cannot recall the right answer—the valid knowledge for that classroom.

Again, though, there is a possible contradiction, because, according to the teacher: "They work together. They help each other but they are also pushing forward." This may open up a way back from exclusion for Joe.

Power is productive. The teacher's is the only voice as she acts out her part. Perhaps it would be claimed that this is inevitable in a video demonstrating "effective teaching." This presents a clear description of teaching and an arguable disassociation of learning from the acts of teaching. The teacher can occupy her position by meeting the description of "effective teaching." For example, the teaching should be pacey. It is then presumed—in fact defined—that the learning is good. For myself, my judgment is that the learning is absent from the scene, replaced by short, sharp activities that keep the children busy and on their toes. It may be difficult for the teacher to become aware of this particular view. It would disrupt what she knows of herself—that she is an effective and caring teacher. This "blindness" reflects the lack of self-awareness that is needed for her self-regulation.

> [Two assumptions] underlie the teaching-learning process: teacher as technician and teacher as therapist [referring to Petrie, 1990, p. 17]. In both cases, a closely scripted strategy of diagnostic-prescriptive moves are legitimated and framed in the language of the human sciences. If the work of teachers (and the research they draw upon) remains at the level of optimizing efficient performances and restoring to health those in their charge, it reduces significantly what it is possible to do to outcomes that can be anticipated and calibrated. It risks leaving teachers unable—indeed, unwilling, to examine what they do and how they speak about what they do within a broader sociopolitical context. Put differently, the privileging of teachers' work along the lines Petrie delineates severely restricts any notion of teacher agency that argues for an open, dynamic, engaged course of action. (Pignatelli, 1993, pp. 419–420)

There is also a fracture between the given description of what constitutes effective teaching and description of a successful lesson where 80% of children learn something new—that is the recognition of the "non-learning" of 20% of the class. A trick of totalization hides the abnormal 20% so the teacher evades the need to question her teaching—a potentially risky business. This reveals the working of the discursive practices. Knowledge is constructed. The subjects of school mathematics, effective teaching, effective learning are all produced here, together with effective teachers and normal learners (e.g., good mathematics tasks are short and change fast. Sustained effort is not desirable. Children do not learn through longer tasks. What they do is just struggle). And the meaning of teachers' and children's actions is oriented around this. The teacher's articulation of herself as teacher points again to the effectiveness of self-regulation. For her it can be a reaffirming process—she can see herself as an effective and caring teacher.

CONCLUDING REMARKS

I have given an example of the effects that our theories and practices can have on the learners of mathematics. These effects are evident not just at the level of learners' actions. They work at a deeper level in forming learners' subjectivity.

> Educators, for example, develop theories about how best to teach and control students, but they also develop instruments and justifications which are used in determining what students can do and, eventually, what they think of themselves. Individuals come to be defined (and self defined) in terms of their "distance" from definitional norms. (Selman, 1989, p. 319)

I have argued that it is important that mathematics education researchers become aware of why the process of learning school mathematics has the effects it does on groups of our children. To achieve this awareness, we need to examine critically our theorizations of the concepts and objects of our study. I have written about my search for more productive ways of examining my practices and discussed how valuable I have found theoretical frames that acknowledge the discursive nature of the field in which I operate.

Following this discussion of the nature of tools or strategies that can appropriately be applied to classroom discourse, I have shown how one such strategy can reveal the discursive practices of mathematics education and the ways that these can position people within the classroom, affecting their actions and determining what each has to say in order to be heard. In particular I have readily recognized these "techniques of power," even in a brief extract of mathematics pedagogic discourse.

I have worked with Foucault's vision of the normative nature of assumption: that, for the vast majority of learners, school is an "unreal" world; where they are not the "normal." The tools I have used were chosen to enable me, as mathematics educator and as researcher, to rethink curriculum and teaching practices, to undermine common sense ideas about children, teachers and mathematics and to gain some insight into why we are not helping many groups of learners learn mathematics.

> Foucault does want to preserve the possibility of agency and choosing to be otherwise, to moving against a life constructed through, and regulated by, a normalizing mode of discourse-practice. But in the face of a form of governing that remains shrouded in the naturalistic garb of the everyday and stubbornly invisible, he also wants us to be aware of what is at stake if we choose to remain silent and inattentive. (Pignatelli, 1993, p. 419)

I will conclude with an indication of two further aspects that might be included in *choosing to do otherwise* as a teacher and as a researcher.

First, for teachers: the video extract used in this chapter is part of a current portrayal of mathematics teachers' practices by UK government agencies. The training videos present exemplars of "effective teaching" as part of teachers' initiation into the National Numeracy Strategy. These exemplars stand for a desired "teacher" and her practices. They invite generalization, by the teachers who have viewed them, to their own and other classrooms and to their sense of their own professional work. In my experience when teachers watch video depictions of mathematics classrooms they often dissociate themselves, their own practices and contexts from such scenarios. In their denials (or acceptance) of the applicability of the portrayed teaching strategies to their own context they are in fact "forgetting" the deliberate production of these images. This evocation of genuine teaching and teachers is one of the ways in which "exemplars" operate and exercise power and produce consent to the "effective teaching" that is generated. This makes it difficult for teachers to question the effectiveness of the teaching methods that are portrayed. This can mean that attempts to resist the normalizing effects of these prescribed practices on both learners and teachers are hard even to consider. Reiterating Pignatelli in the quotation above, it is important *to be aware of what is at stake if we choose to remain silent and inattentive.* This prompts the need for a sustained engagement with questions about encouraging teachers to choose to be otherwise.

Second, for a researcher: in the process of my research I am aware of difficulties in my portrayal and analysis of teachers and their (clearly social) practices from such highly edited and selective sources. In the process of interpreting these extracts, the constructed and political nature of this material can be easily forgotten. I notice how easy it is to slide into asserting something about the (real) teacher who is acting out some (valid) mathematics teaching activity. There is a desire to assert the teacher's motives and the effects of her teaching, and to seek for generalizable claims for my own and other teachers' practices. If I do not do this, I will fail to say or learn anything about mathematics education. However I need to be careful when I write about my results from analyzing video exemplars as discursive text. I can never know what the production of a classroom extract involved. This video has been derived from the activities of teachers, children, classrooms and schools. I do not know what permissions they have given to the production agencies or what they would grant to me. What acknowledgment of this should I and can I give? In the act of interpreting and analyzing the video footage I treat the data as text. I need to continue to ask, "What are the effects of such research strategies?"

In the search for an explanation of the persistence of inequality and the inconsequential changes resulting from reform, I have used the results of a particular analytical technique to respond to the need to become aware of and resist normalizing assumptions about learners, teachers and understandings of mathematics education, and also to consider my own collusion in those normalizing assumptions. This is an uncompleted task. Not least because working with Foucault's constructs of discursive practices and subjectivities reveals the arbitrary and the unstable in my own professional subject knowledge. The products of my own research are contingent and contestable, though I work with extreme care to be able to claim validity and value in my findings. But, there will always be areas where I struggle to consider the effects of the generalities I present.

However, my work starts to reveal and resist the current patterns of normalizing. I suggest that there are ways that mathematics education practices can be changed to disrupt persistence in patterns of inequality, although I do not outline what these might be here. My work also promotes a disruption of assumptions of mathematics education research practices. By working through alternatives, by exploiting the lack of stability of many of our professional notions, we might open up spaces from which we can counter ill-posed problems and look for sites of resistance.

REFERENCES

Appelbaum, P. (1995). *Popular culture, educational discourse, and mathematics*. Albany: SUNY Press.

Askew, M. (1996). Using and applying mathematics in schools: Reading the texts. In D. Johnson & A. Millett (Eds.), *Implementing the mathematics National Curriculum: Policy, politics and practice* (pp. 99–112). London: Paul Chapman.

Ball, S.J. (Ed.). (1990). *Foucault and education: Disciplines and knowledge*. London: Routledge.

Brown, M., Askew, M., Baker, D., Denvir, H., & Millett, A. (1998). Is the National Numeracy Strategy research based? *British Journal of Educational Studies, 46*(4), 362–385.

Bourdieu, P., & Passeron, J-C. (1977). *Reproduction in education, society and culture*. London: Sage.

Department for Education. (1995). *Mathematics in the national curriculum*. London: HMSO.

Department for Education and Employment. (1999). *National Numeracy Strategy: Framework for teaching mathematics from reception to year 6*. Sudbury: DfEE Publications.

Dreyfus, H.L., & Rabinow, P. (Eds.) (1982). *Michel Foucault: Beyond structuralism and hermeneutics*. Chicago: University of Chicago Press.

Dunne, M. (1999). Positioned neutrality: Mathematics teachers and the cultural politics of their classrooms. *Educational Review, 51*(2), 117–128.

Earl, L., Levin, B., Leithwood, K., Fullan, M., & Watson, N. (2001). *Watching and learning 2: Evaluation of the implementation of the national literacy and numeracy strategies.* Toronto: Ontario Institute for Studies in Education, University of Toronto.

Foucault, M. (1972). *The archaeology of knowledge* (Trans: A. Sheridan). New York: Pantheon.

Foucault, M. (1977). *Discipline and punish: The birth of the prison* (Trans: A. Sheridan). New York: Pantheon.

Foucault, M., & Deleuze, G. (1972). Intellectuals and power: A conversation between Michel Foucault and Gilles Deleuze. In D. Bouchard (Ed.), *Language, counter-memory, practice: Selected essays and interviews* (Trans: D. Bouchard & S. Simon, 1997) (pp. 218–233). New York: Cornell University Press.

Hardy, T. (1996). *An investigation into the power relations between mathematics teachers and mathematics curriculum descriptors.* Unpublished Masters Thesis, Manchester Metropolitan University, U.K.

Hardy, T. (2000). Thinking about discursive practices of teachers and children in a "National Numeracy Strategy" lesson. In T. Nakahara & M. Koyama (Eds.), *Proceedings of the 24th Conference of International Group for the Psychology of Mathematics Education* (Vol 3, pp. 33–40). Hiroshima: University of Hiroshima, Japan.

Hardy, T., & Cotton, T. (2000). Problematising culture and discourse for maths education research: Tools for research. In J.F. Matos & M. Santos (Eds.), *Proceedings of the Second International Mathematics Education and Society Conference* (pp. 275–289). Lisbon: Centro de Investigação em Educação da Faculdade de Ciências, Universidade de Lisboa.

Jones, L., & Brown, T. (1999). Tales of disturbance and unsettlement: Incorporating and enacting deconstruction with the purpose of challenging aspects of pedagogy in the nursery classroom. *Teachers and Teaching: Theory and Practice,* 5(2), 187–202.

Klein, M. (1998). How teacher subjectivity in teaching mathematics-as-usual disenfranchises students. In T. Cotton & P. Gates (Eds.), *Proceedings for the First International Mathematics Education and Society Conference* (pp. 240–247). Nottingham: University of Nottingham Press.

Klein, M. (1999). The construction of agency in mathematics teacher education and development programs. *Mathematics Teacher Education and Development,* 1(1), 84–94.

Lerman, S. (1998). A moment in the zoom of a lens: Towards a discursive psychology of mathematics teaching and learning. In A. Olivier & K. Newstead (Eds.), *Proceedings of the 22nd Conference of the International Group for the Psychology of Mathematics Education* (Vol.1, pp. 66–84). Stellenbosch, RSA: University of Stellenbosch.

National Numeracy Project/Hamilton Maths Project. (1998). *Numeracy in action: Effective strategies for teaching numeracy,* Video recording.

Petrie, H.G. (1990). Reflections on the second wave of reform: Restructuring the teaching profession. In S.L. Jacobson & J.A. Conway (Eds.), *Educational leadership in an age of reform* (pp. 17–27). New York: Longman.

Pignatelli, F. (1993). What can I do? Foucault on freedom and the question of teacher agency. *Educational Theory, 43*(4), 411–432.

Rabinow, P. (Ed.) (1986). *The Foucault reader.* London: Peregrine.

Roth, J. (1992). Of what help is he? A review of Foucault and education. *American Educational Research Journal, 29*(4), 683–694.

Selman, M. (1989). Dangerous ideas in Foucault and Wittgenstein. In J. Giarello (Ed.), *Philosophy of Education 1988* (p. 319–329). Normal, Ill.: Philosophy of Education Society.

Valero, P., & Vithal, R. (1998). Research methods of the "north" revisited from the "south'. In A. Olivier & K. Newstead (Eds.), *Proceedings of the 22nd Conference of the International Group for the Psychology of Mathematics Education* (Vol. 4, pp. 153–160). Stellenbosch, RSA: University of Stellenbosch.

Walkerdine, V. (1994). Reasoning in a post-modern age. In P. Ernest (Ed.), *Mathematics, education and philosophy: An international perspective* (pp. 61–75). London: Falmer Press.

Zevenbergen, R. (1996). Constructivism as a liberal bourgeois discourse. *Educational Studies in Mathematics, 31*(1–2), 95–1.

CHAPTER 7

THE PEDAGOGICAL RELATION IN POSTMODERN TIMES

Learning with Lacan

Margaret Walshaw

ABSTRACT

Many explanations about how people learn have been put forward and recent thinking within the discipline has offered valuable insights. I explore some of those ideas and then suggest that the critical work of the postmodernists offers theoretical and empirical directions for a productive analysis of learning. Straddling the ground between Foucault and Lacan, and putting power and the unconscious center stage, I look at some unexplained aspects of one student's learning in brief classroom episodes.

INTRODUCTION

This chapter is a pedagogical journey, exploring how we come to learn. I reframe the core question of learning to offer theoretical insights about what

Mathematics Education Within the Postmodern, pages 121–139
Copyright © 2004 by Information Age Publishing
All rights of reproduction in any form reserved.

it is that prompts us to "take up" ideas. Journeys into learning are hardly novel in mathematics education. But what is different about such journeys nowadays is that they owe some allegiance to scholarly work interrogating modernist ways of thinking. The effects of these critiques for our discipline are still being worked through but what we can say at this juncture is that the focus of our work has sharpened and new analytical frameworks are presenting a sustained challenge to classic ideas about knowing and being.

In the spaces opened by such critique, like that offered by postmodernists, a number of learning theorists within mathematics education have worked more or less independently on a common problem: they have attempted to produce a rigorous method and a satisfactory explanation to capture the situatedness of human thought. The arguments of two main perspectives vary in metaphor and in appeal, and proceed with different emphases from alternative starting points. Differences abound, yet these differences have resulted in strengths and weaknesses that complement each other, and it could be said that those complementary differences are responsible for some of learning theory's greatest insights.

If learning in these tendencies is not a unified theory, there are at least some common themes. I pick up on these themes to extend an issue beyond the scope of current analyses. My general strategy is to apply the critical work of two postmodern writers and suggest theoretical and empirical directions for an analysis of how we learn. I attempt to straddle the ground between Foucauldian poststructuralism and Lacanian psychoanalysis, and apply useful dimensions of both. Taken together they offer, in my view, the most sophisticated and convincing account of subjectivity, and provide a framework and a language for looking at some unexplained aspects of learning.

The journey begins with a look at how conditions for learning are elaborated by two main theoretical currencies, and goes on to explore how these two mark some important criticisms of each other. Next I outline key ideas from the theorizing of Foucault and of Lacan that I believe provide a productive approach to the question of how we might account for "insight." Going against the grain, I foreground the place of power and the unconscious, focusing on those aspects that work psychically for the learner. I engage in that discussion by short example, revisiting an excerpt from my own research that defied comprehension in an earlier analysis (Walshaw, 1999). Reframing the analysis by magnifying those aspects that structure the desire to learn (Britzman, 1998), I explore the psychoanalytic play of affect and its attachments. Arguably, for the discipline of mathematics education, the discussion makes use of "fringe" ideas, yet these ideas do allow the discussion and debate about learning to continue. I believe they are important because they offer possibilities for envisaging change in people's lives.

THEORIES OF LEARNING

Theoretical approaches to learning offered by mathematics education in postmodern times have sometimes been in contest with one another (see Kirshner, 2002). Given the wide range of philosophies, intents and interests underwriting the explanations offered, we could hardly be surprised. Constructivists, as one of the main body of theorists, have taken the individual as their beginning point. They have started with humanist ideas about the learner and have framed their explanations about the status of the individual's learning around that perspective. For sociocultural theorists, another main theorizing group, an interest in social practices sits alongside an interest in the individual learner. Thus sociocultural theorists begin by developing social and cultural conditions and from there have drawn conclusions about the character and shape of learning.

Constructivist Formulations of Knowing

It was during the 1980s and 1990s that "constructivism" became a key term in mathematics education commentaries and became a shibboleth for tendencies in classroom teaching practice and management, teacher education, and research. Piaget (see Inhelder & Piaget, 1964) is regarded as the founding father of constructivism. His genetic evolutionary theory proceeds from the presupposition that it is indeed possible to structure cognition. Learning for him must be understood as intrapsychic activity. The interest is in describing the internal hindrance to the process of developing schemas for meaningful understanding, and much of Piaget's theory is devoted to sketching an answer to the question of how that hindrance actually takes place. Learning, for him, is necessarily precipitated by disequilibrium or cognitive perturbation. The ensuing process of equilibration involves the construction and restructuring of knowledge and with it, the increased ability through stages, to process in the mind the complex information coming from the outside world.

In post-Piagetian work more emphasis is placed on the learner's active construction of mathematical knowledge. Constructivist approaches consistent with von Glasersfeld's (1996) radical position try "to understand the social through its residence in the mind of the individual" (Anderson, Reder, & Simon, 1997, p. 11). This is an individualist approach to learning in which the external context is linked *yet peripheral to* self-organizing individuals. In the constructivist equation, then, the mind is privileged, while circumstances and conditions are minimized. Consistent with that understanding, the individual remains the principal unit of theoretical analysis (see Carpenter, Fennema, Peterson, & Carey, 1988; Schifter & Simon,

1992) and it is the nature of the student's *developing internal representation* that remains of primary interest (Goldin & Shteingold, 2001). In centralizing the learner, constructivist formulations of knowing retain an implicit foundationalist edge since the construction of knowledge relies on the notion of a pregiven individual, and the characteristics of that individual gain precedence in determining the direction of development.

Sociocultural Formulations of Knowing

In contemporary discussions of learning and how we come to know, there can be few terms more widely encompassing in definition than "sociocultural." What is distinctive about sociocultural formulations of learning is their commitment to the social. They reevaluate those themes of learning that revolve around the idea of "mental acquisition" to suggest instead that learning might in fact be "social" or "cultural," constructed by us in engagement and discussion with others.

Sociocultural perspectives draw their inspiration in part from Vygotskian ideas (Forman, 2003). For Vygotsky (1978), the radical constructivist insistence that knowledge is independent of the situations in which it is acquired and used instantiates a specifically limited epistemology. In contrast to the view of the developing child as a "lone scientist," and a focus on her interior mental processes, Vygotsky proposes that the origins of thought are *entirely social* and that conceptual ideas necessarily develop from the intersubjective to the intrasubjective. His formulation of the social construction of knowledge endeavors to "unify culture, cognition, affect, goals, and needs" (Lerman, 2000, p. 37) by prioritizing shared consciousness, or intersubjectivity. Semiotic mediation theory is proposed to account for intersubjective arrangements and the part those arrangements play in the development of internal controls in the learning process.

The Vygotskian understanding of the prior necessity of social interactions for cognitive behavior has become highly influential over the past twenty years. In much work, however, the take-up of Vygotskian ideas is eclectic: many theorists claim an allegiance to the social but nevertheless draw on terms and concepts frequently associated with radical constructivism to describe how learning takes place. In *social constructivist* interpretations of classroom life, special emphasis is placed on the cultural and sociological processes through which knowledge is formulated (Ernest, 1991). In these perspectives the construction of knowledge still remains the preserve of the individual mind, albeit influenced by social and cultural practices. Even as the importance of the social is acknowledged for knowledge construction, the social functions as a shaper rather than a constitutor of learning.

Situated ideas of learning are more all-encompassing in intent, more relational in mode, and more connectivist in function. Situated theories suggest that learning comes about from ongoing participation within a community (Greeno, 2003). From this realization is developed the idea that thinking, meaning and reasoning are constituted socially—a move which has the tendency to stress the mutually relational effects of the social and individual. What is more obviously distinct about theories of situativity, then, is that they do not stop at an analysis of the social context and/or an analysis of the learner: the learner is shown to enter into complicity with the social on both the theoretical level and on the level of practice. It is difficult to separate the cognitive impulse from the societal force.

This inseparability can be seen in the way situated theorists speak about learning as a "social phenomenon constituted in the world" (Boaler, 2000, p. 5), and "an integral part of generative social practice in the lived-in world" (Lave & Wenger, 1991, p. 35), by which they mean to signal that there is no distinction between the learner and the context within which learning takes place. As Lemke (1997) has put it:

> Our activity, our participation, our "cognition" is always bound up with, codependent with, the participation and activity of Others, be they persons, tools, symbols, processes, or things. (p. 38)

Lave's work might well be called the paradigmatic example of situated knowing. In her series of classic studies (e.g., Lave, 1988, 1996; Lave & Wenger, 1991), that draw in part on the work of post-Vygotskian activity theorists such as Davydov and Radzikhovskii (1985), and Vygotsky's collaborator Leont'ev (1978), she assures us that learning is about increasing participation in particular social practices. There is little doubt of the impact and inspiration of Lave's social practice theory on proposals about mathematics classroom culture (see Adler, 1996; Burton, 1999; Jaworski, 1994; Meira, 1995). The main disagreement in the discipline of mathematics education is over the relevance that situativity theories have for questions of learning in the classroom (see Watson, 1998). What does the theory say about how the learner actually learns? What prompts the desire to learn? Lave (1996) claims that there is no need to elaborate a learning mechanism because mechanisms "disappear into practice; mainly people are becoming kinds of persons" (p. 157).

Mechanistic processes, on the other hand, are central to constructivists' formulations of knowing. Through the idea of interiorization they provide an analysis of cognitive conflict and a framework by which the learning process might be explained. But in granting pride of place to mechanisms within cognitive agents, constructivists stake out pedagogical spaces that tend to slide into essentialism. Situated theorists, in turn, offer insightful and convincing criticisms of the central processor model of the mind but their conceptual grounding of individual learning, difference, and mechanisms tends to be underdeveloped. Thus, important critiques of the other are offered by both perspectives. Their arguments are at variance but they do share a commitment to the characteristic orientation of humanist thought—to the fundamental importance of rational thinking and to the rational conscious thinker.

A commitment to the aims and values associated with the development of rational autonomy has been a political and educational aim of Western societies for some time. Walkerdine (1988) has shown how reason is fundamental to all the concepts and the intellectual resources we use to think them. She has noted too how reason's history is often forgotten. Constructivists' accounts of learning necessarily rely on the rational autonomous learner; so too, do socioculturalists' explanations. Walkerdine (1997) argues that Lave's work is about people demonstrating "those characteristics associated with civilized rationality" (p. 60) in different practices. Those characteristics occasion certain prejudices, hierarchies and exclusions, including, as Britzman (1998) points out, the idea that experience is always conscious. Conflating experience with consciousness tends to elevate the coherent and cohesive subject—secure and certain of her own conscious existence. In this arrangement the emotive and unconscious aspects of learning become subordinated and named as irrationalities, and Other to reason. They are deemed intrusions or irrelevancies within pedagogical encounters (Jones, 1996; Walkerdine, 1997).

Walkerdine, Britzman and others (e.g., Bracher, 1999; Ellsworth, 1997; Evans, 2000; Felman, 1987; Jagodzinski, 2002) maintain that the reduction of knowing to conscious experience covers over the complexity in which knowers find themselves. Attempting to get to grips with that complexity sparks a different look at how we come to know and learn. A different look would involve quite different intellectual commitments and demands and a different stance toward pedagogical arrangements. In my mind, Foucault and Lacan provide productive theoretical resources for exploring complexity within the pedagogical encounter.

FOUCAULT

In my exploration of classroom life, cognitive agency owes a large debt to Foucault's ideas about the subject and power. Through all the complexity and scope of Foucault's scholarship, those two notions—the subject and power—are enduring thematic props. It is through these two themes that Foucault is able to unmask rational thinking as intimately tied to the social organization of power. Rational thinking, according to his interrogation, is *not* a quality or attribute of the mind, but rather the result of political struggles between competing perspectives. His major breakthrough comes with the couplet power-knowledge (Foucault, 1980). This is his epistemological signature (Wolin, 1992) and he uses it to convey the idea that power and knowledge directly imply (but are not coextensive with) one another—that there can be no power relation without the correlative constitution of a field of knowledge, nor any knowledge that does not simultaneously presuppose and constitute power relations (Walshaw, 2001).

It is these ideas about power and knowledge that come into play in discussions about the subject. Power, for Foucault, is not merely what we oppose but what we "depend on for our existence and what we harbor and preserve in the beings that we are" (Butler, 1997, p. 2). For Foucault (1972) the human subject as a core rational being cannot provide the foundation for theoretical speculation. One of Foucault's legacies to learning theory was to produce an account of how the subject is produced and regulated in discourses. Certainly Foucault does suggest that power circulates in practices and is capillary in its operation and he has shown how power functions through microlinkages within the social body in everyday social practices. But describing how the subject is produced is not the same as subjectivity—the condition of being a subject. For all his attention to nuance and ambivalence, Foucault stops short of suggesting how the subject is actually enacted into being by power. His account does not explain how power insinuates itself to make us susceptible to new ideas: to attach ourselves to or ignore specific notions and people.

In endeavoring to understand how this process operates in relation to the "take up" of new ideas, Butler shifts the privileges accorded to the mind's conscious activity onto its unconscious modes of operating. She asks: "What is the psychic form that power takes?" (p. 2). Understanding how the psyche contributes to the formation of the subject requires a language for the unconscious, psychical representations and the processes involved in the subject's subjection and receptiveness to ideas. A theory of the psyche will allow us to explore the trajectory of the unconscious and the part it plays in learning. It will enable us to investigate what it is that structures the pedagogical within the mathematics classroom.

LACAN

Psychoanalysis presents complex and well-developed theories of subjectivity. Arguably, psychoanalysis has many shortcomings, yet it does provide us with the most promising theories of how the subject is at once fictional and real. Psychoanalytic theories provide instructive lessons about knowledge and enable us to understand the way power operates to enact the self into being. Grosz (1995) maintains that Lacan's work, in particular, provides a "wide-ranging, philosophically sustained, incisive, and self-critical" (p. 191) account of subjectivity. His radical interrogations address the structure of the psyche, delving into terrain uncharted by Enlightenment thought. They map impossible spaces, confusions and paradoxes, revealing a family resemblance to, for example, the Möbius strip, Escher objects, fractal spaces, and the "strange attractors" of chaos theory (Jagodzinski, 2002).

Lacan builds on Freud's account of psychological development. For Freud, development is not an innate or natural progression through stages, but traces an intrapsychic, interpersonal and social-historical movement. Freud's work, and Lacan's trafficking with it, raise some new questions for understanding the pedagogical relation, learning, and the limits of classroom practice. There are, however, a couple of provisos: first, Lacan's writing is generally regarded as "impossible" (Aoki, 2000, p. 349), "stretching terms to the limits of coherence" (Grosz, 1995, p. 17). And second, there is the question of education itself. As Freud (1937) sees it, education is, along with analysis and government, an impossible enterprise. For all the obscurity of Lacan and all the caution of Freud, the pedagogical potential of psychoanalytic theories, based as they are on an ethic of care, should not be overlooked.

The question I am pursuing with the help of Lacanian terms is the structure of learning. What is it that structures the learning process? Central to an exploration of that question are, in turn, Lacan's development of subjectivity, and the importance of desire and language in that development, his formulation of the three psychic registers, and the place of the unconscious in all of these. I shall attempt to map out the relevance of these terms for a theory of learning.

Subjectivity

Subjectivity, for Lacan, is not constituted by consciousness. In the Lacanian assessment, conscious subjectivity is fraught and precarious and is clearly not the coherent and authentic source of the interpretation of reality as assumed by constructivist and situated discourses of knowing. The psychoanalytic criterion of subjectivity does not rely on a pre-given and self-transparent subject who is in control of her own thought, but is heavily dependent on one whose ontological status remains permanently unclear. That status is constantly under threat precisely because consciousness is continually subverted by unconscious processes. What this means for the learner is that consciousness is not the strategic organizer of her intentions or her learning experience. Unconscious processes will always interfere with conscious intentionality and experience (Britzman, 1998).

It is Lacan's explanation about how the unconscious is structured and the implications of that structuring for the discourses of consciousness, which have major significance for psychoanalysis. Lacan uses a theory of signification, appropriating and modifying Saussure's structuralist analysis of signs, to explain how the unconscious works, and to establish a connection between words and ideas. In this well-known work he considers the unconscious as a site of repression and elusive meanings, made up of signifiers unable to access consciousness. For Lacan, the unconscious consists of a language whose ideas are not known through linear progression, stages of development, singular chronology, and mastery but through ambivalence, discontinuities, diversions and continual openness (Britzman, 1998). The unconscious is the place where, for example, we can hold together two opposite understandings. As Britzman (2001) has noted, it is what holds thinking back, and races it forward, or causes us to lose or forget our focus. The subject, in all this, is spoken as a subject through the unconscious. The subject cannot be conceived as the master of discourse, but "is the effect of discourse, no longer its cause" (Grosz, 1995, p. 98).

The fundamental Lacanian principle that the subject is always "already rhetorically marked" (Guerra, 2002, p. 7) points to a different set of presuppositions from those central to constructivist and situated theories of knowing. In these latter theories, just as in more traditional views, cognitive know-how rests upon the modernist conception of the conscious and rational knower. Lacan's theory of the psyche, as a major change in thinking about how we come to be and how we come to think, seems to suggest that there are thoughts already in place in the mind even before we think them. If there is more going on in the learner's mind than conscious rational thinking it would be useful to know how a thought or a teacher's lesson attaches itself to a thinker.

Desire and Language

Lacan confronts the question of attachment with the dynamics of the emotions. Colliding epistemology and ontology, he maintains that the subject's very existence consists of desire. Desire is the motor of human action, a motivator that lies at the core of the subject and hence is fundamental to the trajectory of her life. As a "positive production, [desire is] the energy that creates things, makes alliances, and forges interactions between things" (Grosz, 1994, p. 75). Desire in the Lacanian analysis looks very like the Nietzschean realization of the will-to-power. What Lacan is able to do is theorize of what that will-to-power consists.

Marked by both conscious and unconscious intentionality, desire "begins to take shape in the margin in which demand becomes separated from need" (Lacan, 1977a, p. 311). As the "*reality* of the unconscious" (Grosz, 1995, p. 67), language plays a key role in its dynamics. But as an effect of language desire is ambiguous, "an element necessarily lacking, unsatisfied, impossible, misconstrued" (Lacan, 1977b, p. 154). Desire, in Lacanian analysis, is not about conquest and attainment, but first and foremost, about the quest for a secure identity. As in all aspects of everyday life in which the learner finds herself, the desire for secure identity is dependent on a desire for the other's recognition. "Desire desires the desire of another" (Grosz, p. 65) and it is this activity that allows consciousness to turn itself into self-consciousness. The learner in the classroom could not be that person without relationships, location, networks and history that allow her to fabricate a presence of self-coherence and effectivity. The desire for self-presence, however, will always be subjected to the constant deferral of satisfaction.

Three Psychic Registers

To understand the place of the desiring learner, Bracher (2002) and Schlender (2002) suggest that we look at Lacan's tripartite register of the subject and identity formation. Lacan formulates three psychic registers of subjectivity—the Symbolic, the Imaginary, and the Real. They function interdependently, working together to inform the learner's experience and sense of perception. Each is responsible for processing its own set of "data," namely, "affects, percepts, or concepts" (Bracher, 2002, p. 99), and it is up to the learner to "make peace" with the conflict among the forms of recognition that each offers.

The symbolic for Lacan is the domain of laws, words, letters, and numbers that structure our institutions and cultures. The Symbolic register accounts for the learner's perception of and relation to learning in a particular classroom. What the symbolic allows (and what it disallows) is derived from the "laws" of the larger social order, or in Lacanian terminology, the "Law of the Father" and the "Big Other." For example, in my investigation of the constitution of female subjectivity in a coeducational classroom (Walshaw, 1999), the Big Other included the official curriculum statement, the rules and procedures of the school community, and the norms of the classroom. And one thing that the research school community validated was the "senior academic mathematics student." Simply being enrolled in the academic class in which the study took place meant recognition from family, friends and other students. Within the classroom itself other representations, meanings and significances were at play. In particular, students were expected to share and justify their ideas and were not permitted any attention seeking. The students of this class desired recognition from each other and from their teacher, as they worked at embodying those signifiers, continuously interplaying the presence and absence of them within themselves. When they succeeded, the recognition became a motivator and learning was effective. A failure to be recognized, on the other hand, contributed to many of the small-scale personal dramas in this classroom.

Lacan's Imaginary register is the realm of visual-spatial images and illusions of self and world. It lies at the limits of perception. The Imaginary order is produced from the conflict between perception and misrecognition that occurs initially at the "mirror stage" when the infant's first image of itself in a mirror is split from his or her self-perception (Grosz, 1995). Through the absences and alienations it conveys the Imaginary register undermines the individual's sense of self. We can grasp a sense of what Lacan means by looking at how the Imaginary order operates in the secondary school classroom. In the study referred to earlier, many female students worked hard to construct a sense of self and bodily appearance which was not simply an illusion of the sort of image over which they fantasized. Field notes taken from observations in the research classroom suggested that the desire for Imaginary register recognition played out as in-class attention to posturing, grooming, and the like, but did not appear to disrupt learning. In Lacanian thinking, a minor conflict between the "contents" of the Imaginary and Symbolic registers became of no consequence.

Lacan's Real Register is not the same as reality. It is elusive, and "defies all description and representation" (Jagodzinski, p. xii). We might think of it as an extradiscursive site for all things that the Symbolic and Imaginary registers cannot contain. The Real register points to a "lack of a lack" (Lacan, 1977a, p. 55); it is an indicator of our sociopsychical growth. Desire for recognition in the Real register is concerned with the mirroring of affect and emotion. Pedagogically the desire for emotional resonance can be extremely productive (Bracher, 2002), as when, for example, the learner and the teacher are both caught up in the same positive intensity for particular mathematics exploration or conceptual understanding. What is it that makes the learner want to mirror the teacher's desire? The desire escapes expression but can be construed as a range of impressions and feelings that emerge "from the holes, pauses, and cavities while perception awaits the achieving of meaning" (Jagodzinki, 2002, p. xvi). Those impressions and feelings pass through memories and unconscious desires and can be triggered by, among other things, a gesture, or the tone, pitch, or resonance of the teacher's voice (Britzman, 2001).

In the next section I explore these three psychic registers in operation. I use the registers and the information that each processes, together with Foucault's ideas about power-knowledge, to look briefly at how one student, Emma, constructed herself as a learner in my study on gendered subjectivity.

THE SILENT ENGAGEMENT OF MATHEMATICAL WORK

Emma is a Year 12 student at a large metropolitan secondary school. Emma is not an extension student; nor was she seen to be struggling with mathematics. According to her teacher she was well behaved, and produced neat and tidy work. "I'm a quiet sort of person. I enjoy doing new things and stuff, but I'm not really that outgoing" [interview transcript]. She appears reluctant to contribute to or to ask questions in whole class discussions. My hope was that I might yet be able to identify what sense of the Real she constructs to legitimate her classroom efforts. What images of learning worked through her practices? How did those images match the "laws" of the classroom? This was how Emma describes her mathematical practice to me:

> During the lesson I usually listen and when she's [the teacher] finished, I write everything down, what she's done. Otherwise I'm writing and listening at the same time and it's sort of … , it's not really … , I don't really understand it totally. [I] just watch. Like, if she's writing while she's explaining, I watch what she's doing and when you're … Write it down. And when you're by yourself, just go through it again. Don't just copy it straight down; just take your time writing it. Sometimes it might mean that you're behind someone else who has been busy writing it down as the teacher has been talking. But that doesn't matter.

Emma orchestrates a relationship between mathematics and the student in which the learner is to "watch," then "write it down" and later "go through it again." In this logic, one moves reflexively from the teacher's talk to the writing and back to the remembered talk, grounding interpretation through the process of writing. In this way, teacher talk, the learner, and mathematical knowledge are linked.

Within the structure of Emma's talk lies an implicit understanding of how to conduct pedagogical relations in the classroom. I was interested in how she enacted those relations. In analyzing Emma's engagement with learning the crucial point to note is that though Emma rarely speaks to her peers, unlike those same peers she frequently seeks assistance from her teacher. Within the sociocultural perspectives that organize the practices of this school's mathematics classrooms, Emma's practice would be understood as pathological. She is passive and does not engage powerfully because she does not speak or reveal mathematical "truths"; nor does she engage the "correct" procedures to produce those "truths" with her peers.

To the classroom Emma presents the face of the hard-working diligent student, in which, while pathologized through the Symbolic order of this classroom, she is still desired. From an informal discussion with Emma's teacher it was clear that a long history of teaching (almost thirty years) had helped form in Mrs. Southee (the teacher) certain images of what a "model student" might look like. These images are at odds with what teachers in this school were expected (and endeavored) to put into practice. In reading the conflicting Imaginary impressions and enacted practices of her teacher correctly, Emma has worked out what it means to be empowered for learning in this classroom. And, in return, it is through reference to Emma's quiet, well-behaved manner that the teacher is able to justify the attention she gives to her: she is seen to be deserving.

The information processed by the Imaginary and Symbolic registers, and the way Emma acts upon this, places her in a very powerful position. She can receive her teacher's approving gaze by being silent with her peers. Simultaneously, she has also learned that she can prompt this attention by positioning herself as often in need of mathematical support and help—not with the risk of whole class attention drawn toward herself, but in the close teacher-student one-to-one interchange. In this way she is able to sustain her productive learning position. It is not difficult to see why "asking for help" might mean something for Emma at more than the rational level; how it might mean something to her psychically. She desires the desire of the teacher. She desires emotional resonance. For her, this desire is pedagogically advantageous.

I have selected a few of many such occasions to illustrate what takes place:

> **Emma**: Mrs. Southee? What's "c?" Do you just have to work out... ? [22 July, 1998, 2.03pm]
>
> **Emma**: Mrs. Southee? How do you know the distance from the aeroplane? [22 July, 2.06pm]
>
> **Emma**: Where's Mrs. Southee? Where's she gone? I'll ask her when she comes back. [28 July]
>
> **Emma**: Mrs. Southee? Is this exactly the same with a third instead of a half? [29 July]
>
> **Emma**: Do they look right there? [*to Mrs. Southee*] [12 August, 9.13am]
>
> **Emma**: I'm confused! I'll ask Mrs. Southee. Mrs. Southee? [12 August, 9.42am

> **Emma:** Mrs. Southee? Can you help me with number twelve? [13 August, 10.55am]
>
> **Emma:** Mrs. Southee? I'm not sure about drawing the line on them. [13 August, 11.07am]

Emma has worked out an answer to be $x^{7/3}$. She compares this with the answer given in the book as $\sqrt[3]{x^7}$.

> **Emma:** Mrs. Southee? Why do they change that one to the square root?
>
> **Mrs. Southee:** No, it wouldn't be. It would be the CUBE root of x to the seven. Because it's given in that form.
>
> **Emma:** So you have to give it back in that form? So that's the same thing? [7 August]

In an earlier lesson Emma has revealed subtle confusion over the signs of exponents:

> **Emma:** Mrs. Southee? Why, for number two, all of the answers you've changed that, and put the x to the power whatever?
>
> **Mrs. Southee:** Well, that's negative on top and it becomes positive on the bottom …
>
> **Emma:** But why is that one, you've written it as x to the power of five?
>
> **Mrs. Southee:** Well, it's already positive.
>
> **Emma:** So if it's negative up there …
>
> **Mrs. Southee:** It is convention to write things with positive exponents. [30 July]

Mrs. Southee uses Emma's confusion as indicative of problems that all her students might be experiencing. Immediately after speaking with Emma she returns to the board to speak to the whole class. This sequential practice of helping Emma, followed by whole-class clarification, is enacted frequently. Emma is instrumental in foregrounding mathematical problems common to all students. In my analysis, the emotional mirroring takes place and Emma's identification of herself as a mathematics learner in this class becomes (provisionally) secure.

> **Mrs. Southee:** [*to class*] X to the power negative two is one over x squared. Are you all happy about this? If you have one over "a" to the power negative three, it's the same as "a" cubed. If you have one over "m" to the power four, it's going to be "m" to the power negative four. Now it's convention, listening?, it's convention to write things with positive exponents … [30 July]

The knowledge-power relations are clear. Emma constructs for Mrs. Southee a version of the student, and a version of teaching-learning, just as her subjectivity constructs a pedagogical position for Mrs. Southee. In that Emma's mathematically divergent propositions are not discounted but are instead used as a benchmark to reproduce and naturalize other more conventionalized forms of mathematical logic, Emma contributes to the symbolic processes of what is to count as mathematical knowledge within the classroom. Thus mathematical knowledge, in this instance, is inseparable from (yet not reduced to) the psychic power of learning.

LEARNING THROUGH SUSTAINING THE ILLUSION OF SECURE IDENTITY

Emma's work is a practice of inventive, resourceful and strategic moves in which desire for recognition plays a crucial part. Role modeling is not at stake here because her work exceeds the mimetic role offered to the student in role model pedagogies. Rather, her work is formed in the gaps, the silences, and the contradictions within and among classroom social practices and her images of them. It is through her repeated performances of familiar strategies of self construction (Britzman, 2001) that we can focus on how she creates herself as a learner, in connection with others. For Emma's work in the mathematics classroom, these performances operate to sustain her illusion of a coherent and cohesive self.

Emma questions her own limits of knowledge, aware of the irreconcilable tension between the search for an "empowering" place from which to speak, and within which to act, and the price at which this secure place will need to be bought. She may not yet be able to articulate that price. She locates spaces where possibilities of advantage for herself in this classroom might arise. The idea, then, of Emma's search for a secure identity refers to those strategic projects by which, through resolving conflict between psychical registers, she sets herself rules of conduct, and by which she intentionally works at optimizing both her classroom practice and secure identity. There is no struggle over meanings between Emma and her teacher about what it means to be a learner in this classroom. In that respect the classroom becomes a safe and secure place in which Emma might speak and act. Inevitably that secure identification will produce new knowledge for her. In producing herself in this way she is involved in an ongoing and complex interplay with knowledge and its attachment to her. She is involved in learning.

CONCLUSION

This chapter marks a pedagogical journey into insight. It offers an explicit personal account of changes in my thinking about learning over a short period of time. It traces an engagement with questions of how learning takes place in constructivist and sociocultural classrooms with my current re-engagement with questions of subjectivity and knowledge. This re-engagement disrupts and challenges assumptions about knowledge and subjectivity upon which mathematics education has been developed. Challenging assumptions and problematizing universals are the very tasks that are integral to the postmodern project.

I have outlined some fundamental concepts from Foucault's work and from Lacanian theory that enable us to reframe central issues and problems about learning. Foucault provides an understanding of power and knowledge as reciprocally constitutive; Lacan is able to theorize how that constitution takes place. The choice of the Foucauldian and psychoanalytic concepts has been deliberately selective, to allow us to repunctuate an analysis of subjectivity with an account of subject positions produced through power-knowledge relations. Drawing on these ideas I suggest theoretical directions for thinking about learning as a psychic event.

Taking a tentative first step to address the problems associated with the construction of the conscious rational learner, I looked at how Emma lived her classroom practice and how she operated strategically in the limited terrain of self-production which was open to her. Using psychoanalytic theory I tried to make sense of my classroom observations and the audio transcripts that appeared to question her mathematical engagement. The analysis revealed the way in which her silence might be mistakenly read as pathological; in doing so it pointed to the contradictions, hidden and distorted by current understandings of learners in school mathematics. For Emma, what is fundamental to learning is her desire for the teacher's desire; it is what attaches the psychical to classroom practice, and classroom practice to the psychical. She reveals an astute management in securing that desire. Securing the emotional resonance she desires is precisely what enables a mathematical idea to attach itself to her, and allows her to advance and flourish—to learn productively—in the mathematics classroom.

Emma's identification with learning is marked, albeit always provisionally, by the place where the "inside" or the psychic, meets the "outside" or the social. The transference of knowledge becomes a question, not about conscious experience with self and others, but rather to do with the way in which unconscious processes, working at different levels and with different kinds of information, undermine experiential knowing. The place of the unconscious, then becomes central to the transference of knowledge. Learning turns out to be a mode of activity circumscribed far beyond the rational autonomous knower. It is mediated by unseen, unspoken, atemporal coordinates, that serve to undermine any certain basis for reasoned learning.

REFERENCES

Adler, J. (1996). Lave and Wenger's social practice theory and teaching and learning school mathematics. In L. Puig & A. Gutiérrez (Eds.), *Proceedings of the twentieth Conference of the International Group for the Psychology of Mathematics Education* (Vol. 2, pp. 3–10). Valencia, Spain.

Anderson, J.R., Reder, L.M., & Simon, H.A. (1997). Situative versus cognitive perspectives: Form versus substance. *Educational Researcher, 26*(1), 18–21.

Aoki, D.S. (2000). The thing never speaks for itself: Lacan and the pedagogical politics of clarity. *Harvard Educational Review, 70*(3), 347–369.

Boaler, J. (2000). Introduction: Intracacies of knowledge, practice, and theory. In J. Boaler (Ed.), *Multiple perspectives on mathematics teaching and learning* (pp. 1–17). Westport, CT: Ablex.

Bracher, M. (1999). *The writing cure: Psychoanalysis, composition, and the aims of education.* Carbondale: Southern Illinois University Press.

Bracher, M. (2002). Identity and desire in the classroom. In J. Jagodzinski (Ed.), *Pedagogical desire: Authority, seduction, transference, and the question of ethics* (pp. 93–122). Westport, CT: Bergin & Garvey.

Britzman, D. (1998). *Lost subjects, contested objects: Toward a psychoanalytic inquiry of learning.* New York: State University of New York Press.

Britzman, D. (2001). The arts of inquiry. *Journal of Curriculum Theorizing, 17*(1), 9–26.

Burton, L. (Ed.) (1999). *Learning mathematics: From hierarchies to networks.* London: Falmer Press.

Butler, J. (1997). *The psychic life of power.* Standford, CA: Standford University Press.

Carpenter, T.P., Fennema, E., Peterson, P., & Carey, D. (1988). Teachers' pedagogical content knowledge of students' problem-solving in elementary arithmetic. *Journal for Research in Mathematics Education, 19*, 385–401.

Davydov, V.V., & Radzikhovskii, L.A. (1985). Vygotsky's theory and the activity-oriented approach in psychology. In J.V. Wertsch (Ed.), *Culture, communication, and cognition: Vygotskian perspectives* (pp. 35–65). New York: Cambridge University Press.

Ellsworth, E. (1997). *Teaching positions.* New York: Teachers College Press.

Ernest, P. (1991). *The philosophy of mathematics education.* London: Falmer Press.

Evans, J. (2000). *Adults' mathematical thinking and emotions: A study of numerate practices.* London: Routledge Falmer.

Felman, S. (1987). *Jacques Lacan and the adventure into insight: Psychoanalysis in contemporary culture.* Cambridge, MA: Harvard University Press.

Forman, E.A. (2003). A sociocultural approach to mathematics reform: Speaking, inscribing, and doing mathematics within communities of practice. In J. Kilpatrick, W.G. Martin, & D. Schifter (Eds.), *A research companion to Principles and Standards for School Mathematics* (pp. 333–352). Reston, VA: NCTM.

Foucault, M. (1972). *The archaeology of knowledge and the discourse on language* (Trans: A. Sheridan). New York: Pantheon.

Foucault, M. (1980). *Power/knowledge: Selected interviews and other writings, 1972–1977 Michel Foucault* (Trans: C. Gordon). New York: Pantheon Books.

Freud, S. (1937). Analysis terminable and interminable. In *The standard edition of the complete psychological works of Sigmund Freud* (Trans: J. Strachey) (vol 23, pp. 209–254). London: Hogarth.

Goldin, G., & Shteingold, N. (2001). Systems of representations and the development of mathematical concepts. In A.A. Cuoco & F.R. Curcio (Eds.), *The roles of representation in school mathematics: 2001 Yearbook* (pp. 1–23). Reston, VA: NCTM.

Greeno, J.G. (2003). Situative research relevant to standards for school mathematics. In J. Kilpatrick, W.G. Martin, & D. Schifter (Eds.), *A research companion to Principles and Standards for School Mathematics* (pp. 304–332). Reston, VA: NCTM

Grosz, E. (1994). Refiguring lesbian desire. In L. Doan (Ed.), *The lesbian postmodern* (pp. 67–84). New York: Columbia University.

Grosz, E. (1995). *Jacques Lacan: A feminist introduction*. London: Routledge.

Guerra, G. (2002). Psychoanalysis and education? In J. Jagodzinski (Ed.), *Pedagogical desire: Authority, seduction, transference, and the question of ethics* (pp. 3–10). Westport, CT: Bergin & Garvey.

Inhelder, B., & Piaget, J. (1964). *The early growth of logic in the child*. New York: Routledge & Kegan Paul.

Jagodzinski, J. (Ed.) (2002). Introduction. *Pedagogical desire: Authority, seduction, transference, and the question of ethics*. Westport, CT: Bergin & Garvey.

Jaworski, B. (1994). The social construction of classroom knowledge. In J. da Ponte & J. Matos (Eds.), *Proceedings of the eighteenth Conference of the International Group for the Psychology of Mathematics Education* (pp. 73–80). Lisbon, Portugal.

Jones, A. (1996). Desire, sexual harassment, and pedagogy in the university classroom. *Theory into Practice, 35* (2), 102–109.

Kirshner, D. (2002). Untangling teachers' diverse aspirations for student learning: A crossdisciplinary strategy for relating psychological theory to pedagogical practice. *Journal for Research in Mathematics Education, 33*(1), 46–58.

Lacan, J. (1977a). *Ecrits. A selection*. London: Tavistock.

Lacan, J. (1977b). *The four fundamental concepts of psycho-analysis*. London: The Hogarth Press.

Lave, J. (1988). *Cognition in practice: Mind, mathematics and culture in everyday life*. Cambridge: Cambridge University Press.

Lave, J. (1996). Teaching, as learning, in practice. *Mind, Culture, and Activity, 3,* 149–164.

Lave, J., & Wenger, E. (1991). *Situated learning: Legitimate peripheral participation*. Cambridge: Cambridge University Press.

Lemke, J. (1997). Cognition, context, and learning: A social semiotic perspective. In D. Kirshner & J.A. Whitson (Eds.), *Situated cognition: Social, semiotic, and psychological perspectives* (pp. 37–56). Mahwah, NJ: Lawrence Erlbaum Associates.

Leont'ev, A.N. (1978). *Activity, consciousness, and personality*. Englewood Cliffs, NJ: Prentice-Hall.

Lerman, S. (2000). The social turn in mathematics education research. In J. Boaler (Ed.), *Multiple perspectives on mathematics teaching and learning* (pp. 19–44). Westport, CT: Ablex.

Meira, L. (1995). Mediation by tools in the mathematics classroom. In L. Meira & D. Carreher (Eds.), *Proceedings of the nineteenth Conference of the International Group for the Psychology of Mathematics Education* (pp. 102–111). Recife, Brazil.

Schifter, D., & Simon, M. (1992). Assessing teachers' development of a constructivist view of mathematics learning. *Teaching and Teacher Education, 8,* 187–97.

Schlender, B-A. (2002). Sexual-textual encounters in the high school: "Beyond" reader-response theories. In Jagodzinski, J. (Ed.), *Pedagogical desire: Authority, seduction, transference, and the question of ethics* (pp. 123–146). Westport, CT: Bergin & Garvey.

von Glasersfeld, E. (1996). Aspects of radical constructivism and its educational recommendations. In L. P. Steffe & P. Nesher (Eds.), *Theories of mathematical learning* (pp. 307–314). Mahwah, NJ: Lawrence Erlbaum Associates.

Vygotsky. L. (1978). *Mind in society: The development of the higher psychological processes.* Cambridge, MA: Harvard University Press.

Walkerdine, V. (1988). *The mastery of reason: Cognitive development and the production of rationality.* London: Routledge.

Walkerdine, V. (1997). Redefining the subject in situated cognition theory. In D. Kirshner & J.A. Whitson (Eds.), *Situated cognition: Social, semiotic, and psychological perspectives* (pp. 57–70). Mahwah, NJ: Lawrence Erlbaum Associates.

Walshaw, M. (1999). *Paradox, partiality and promise: A politics for girls in school mathematics.* Unpublished doctoral dissertation, Massey University, New Zealand.

Walshaw, M. (2001). A Foucauldian gaze on gender research: What do you do when confronted with the tunnel at the end of the light? *Journal for Research in Mathematics Education, 32*(5), 471–492.

Watson, A. (Ed.) (1998). *Situated cognition and the learning of mathematics.* University of Oxford Department of Educational Studies, UK: Centre for Mathematics Education Research.

Wolin, R. (1992). *The terms of cultural criticism: The Frankfurt school, existentialism, poststructuralism.* New York: Columbia University Press.

CHAPTER 8

AFFECT AND COGNITION IN PEDAGOGICAL TRANSFERENCE

A Lacanian Perspective

Tânia Cabral

ABSTRACT

In this chapter I explore affect and cognition from Lacan's psychoanalytic theory. Employing the term 'pedagogical transference,' I present a view of a psychoanalytically-inspired classroom organization. In analyzing how two students engage with mathematics I show how the mathematical 'nonsense' produced by the student can represent the teacher's object of desire. It is my contention that to accept the psychoanalytical perspective is to accept that the teacher is directly implicated in the cognitive problem.

Mathematics Education Within the Postmodern, pages 141–158
Copyright © 2004 by Information Age Publishing
All rights of reproduction in any form reserved.

INTRODUCTION

Currently researchers in mathematics education are looking beyond traditional theories of how students learn mathematics. Numerous studies are now focusing on what learning might mean for students in terms other than differential performance and some of these have dealt with affect. This chapter offers yet another approach to the study of affect and cognition in pedagogical transference. As one working in mathematics education but trained in psychoanalysis, I argue that the most representative studies dealing with affect in mathematical education miss the interpretation of cognition in terms of transference. I base this claim on my review of a number of papers that focus on feelings of students engaged in mathematical tasks and approach affect either from a common sense view or from a psychoanalytical perspective.

I begin my discussion by noting the ways in which these papers deviate from Lacanian psychoanalysis in terms of affect and how they provide an incomplete description of learning experiences. Next, I introduce Lacan's concept of *transference*: a *phenomenon in speech* where two people become committed to what they say. This makes it possible to introduce the dimension of *love* that had not appeared in the above references. According to Miller (2001), love is the name of the *transferential belief*. I introduce the signifier *pedagogical transference* to refer to a special case of the concept of transference to the mathematics learning/teaching experience.

Two cases of learning experiences are presented; one concerns straightline topology, the other is about uniform convergence. The psychoanalysisinspired classroom organization is called an *integrated session*. As in cognition studies, these sessions focus on the direct interaction of students with a mathematical object. However, the commitments of those involved in the sessions are different—the teacher and the student are constituted repeatedly as they interact through language. The teacher listens, the student talks. The mathematical misconception produced by the student becomes the teacher's object of desire (*petit-a*) as the teacher becomes "hypnotized" by the student's misunderstanding.

Last, I contend that Lacanian psychoanalytic theory has much to offer studies of cognition in mathematics education (Cabral & Baldino, 2002). To accept the psychoanalytical perspective is to accept that the teacher is directly implicated in the cognitive problem and, as such, her behavior cannot be controlled nor taken for granted as an external reference. All the "non-sense" and difficulties that students experience must be accepted as part of the researched unknown.

WHAT A LITERATURE SEARCH HAS REVEALED

Much has been written about affect in mathematics education. However, as Breen (2000) notes in his study of students' fears and anxiety disclosed through writing, "the links between psychoanalysis and mathematics education seem to have largely been silent themes ... with only a few discernible exceptions and neither of these directly address the encountered dominance of fear in the mathematics classroom" (Breen, p. 108). Breen's observation that those learning experiences described in the literature are not routinely organized according to psychoanalytical theory, suggests that it is timely to take a brief look at how they are, in fact, organized.

The Affective Domain as it is Portrayed in Mathematics Education

McLeod's (1992) survey displays an impressive list of 219 literature references that fall under the general heading of affect yet none refer directly to psychoanalysis—Freud is mentioned *en passant* in the article. McLeod says that "the affective domain refers to a wide range of beliefs, feelings and moods that are generally regarded as going beyond the domain of cognition" (p. 576). However "research on affect in mathematics education continues to reside on the periphery of the field" (p. 575), "a major difficulty being that research on affect has not usually been grounded in a strong theoretical foundation" (p. 590). Indeed, from the survey it is surmised that the affective domain spans common sense terms such as anguish, anxiety, attitudes, autonomy, beliefs, confidence, curiosity, dislike, emotions, enthusiasm, fear, feelings, frustration, gender, hostility, interest, intuition, moods, panic, perseverance, sadness, satisfaction, self-concept, self-efficacy, suffering, tension, and worry.

Not much has changed since McLeod's report (see any recent conference proceedings of the discipline). The few attempts that do approach affect issues from the psychoanalytical perspective all evoke the concept of *transference.* Those of most consequence are commentaries from Blanchard-Laville (1997), Wilson (1995), and Breen (2000). Blanchard-Laville (1997) postulates a certain "psychical reality" that could only be known through its effects, while remaining unknown in itself, like Kant's thing-in-itself. She acknowledges the dilemma that in order to study the human psyche we must pass through another human psyche, namely, our own. Blanchard-Laville offers a solution to the dilemma. The solution is a direct appeal to the conscious ego: "A minimum of conscious intentions is required of the observer in order to perceive the unconscious dynamic of exchanges and its reflections and let oneself impress by the implicit aspect of messages among the participants of the didactical exchange" (1997, p. 158). Such an appeal to a self-controlling conscious will, however, is untenable from a Lacanian perspective.

Wilson (1995) studies his own feelings in his pedagogical relationship with students. He describes his method thus: "At the end of each day, or week, I sat quietly and allowed an incident from my teaching to enter my mind ... writing [it] as objectively as I could" (Wilson, 1995, p. 1). To this introspective method he adds the concept of transference which he describes in these terms: "The transference relationship describes distorted perceptions of counselors which arise because of clients' previous relationships" (1995, p. 1). In an approach reminiscent of Blanchard-Laville's, an autonomous conscious subject would be called upon in order to judge which perceptions are genuine and which are distorted.

Breen (2000) reports on his teaching experience with 'second-chance' adults who were seeking a primary school teaching diploma. Transference is considered as "the imposition of an actual or imagined previous relationship onto a present one" (Breen, 2000, p. 110). The focus would thereby be diverted from the actual scene to elsewhere in a supposed objective past.

Of all those learning experiences reported none satisfies a strictly psychoanalytic interpretation. I make this claim on the basis that the focus of these characteristic cognition studies is not on the direct interaction of student or teacher with the mathematical object. Whenever the interaction between the subject and a mathematical object *is* brought about, the focus is on the subject's *feelings*, described from the point of view of a superior conscious ego who relies on introspection to evaluate imaginary distortions of an exacting pattern. It seems researchers strongly believe that psychoanalytic theories and affective studies explain what happens on the periphery of cognition and have nothing to say about cognition itself.

In previous attempts at linking cognition with affect the researcher assumes an exterior position either as an observer and interpreter—as in Vinner (1996), Da Rocha Falcão and Hazin (2001)—or assumes a conscious position of judge—as in Blanchard-Laville (1997) and Wilson (1995). In doing so, the researcher approach is via an "alliance with the healthful part of his [sic] own self" (Lacan, 1973, p. 4) which generally runs in an opposite direction to the unconscious reality that psychoanalysis is expected to actualize.

Schlöglmann (2002), on the other hand, presents a study about affect supported by results of neuroscience. He argues that some categories of affect like emotions, beliefs, attitudes, and values, "are founded on mental background systems which control thinking, learning and acting" (Schlöglmann, p. 185). He recognizes that there are studies that stress the social context and that this is an important concept in research on affect in learning. In his theoretical paper he points out that the framework of neuroscience assures that: first, cognition and emotion are two subsystems of the brain system, and second, that they interact. On the basis of reports of studies on emotions, Schlöglmann promotes the idea that there are two kinds of memory systems: (1) the implicit emotional memory that operates unconsciously and (2) the explicit memory that operates consciously. This second aspect clearly implies that there is a part of the memory that belongs to the cognitive domain and in that domain everything is controlled and known. Cognition and emotion are highly polarized and their interaction must be controlled from a conscious center.

For Freud, the subject has no conscious center. Freud showed us that the subject of the unconscious is the actual master of ourselves. In his clinic he learned with Dora that sometimes we say or do things that we did not intend to say or do (*acte raté*). Those of us who work in psychoanalysis accept that nothing we say and do can be absolutely controlled. The *acte raté* is not the act of an autonomous conscious subject who has absolute control over every word said.

Lacan, in developing Freud's work, offers theoretical resources that we can draw on to assist in understanding the idea of the decentered subject as it relates to learning. In the next section I attempt to show how theory elaborated by Lacan to explain the dialogic encounter with his patients, may be used both to explain and orient the learning/teaching experience in a mathematics classroom. For me, Lacan's theory provides the theoretical resources for dealing with the learning experience.

THE THEORETICAL BACKGROUND

According to Lacan (1977), there are four fundamental concepts of psychoanalysis: the *repetition*, the *unconscious*, the *transference* and the *triebe*, and they should ordinarily not be treated separately. However, here the focus will be on transference only and this particular focus will be considered with *imaginary identification*. It is hoped that what is lost in fidelity might be gained in clarity. According to common sense ideas concerning psychoanalysis, transference is either a substance transmitted between subjects via communication such as the "transmissions psychiques" of Blanchard-Laville (1997), or a catharsis of unconscious elements displaying a distortion to be rectified by counter transference, as in Wilson (1995). Transference is also thought of as an affect that may be positive or negative. Positive transference is identified with love.

Lacan (1973) begins with this common sense conception and takes it elsewhere. According to its meaning in clinical experience, "transference is a phenomenon where both the subject and the psychoanalyst are included together" (p. 210). It needs to be pointed out that this is not a symmetrical phenomenon occurring between two subjects. In fact, in analysis, transference is a particular symbolic experience that guides the analyst toward building up a subjective logic. This particular logic refers to all those things that have constituted the subject of the unconscious. Since, in this sense, transference is "the actualization (*mise en acte*) of the reality of the unconscious" (Lacan, 1973, p. 137), the analytic experience promotes the experience of desire.

What has to be analyzed in the clinic is transference itself. "According to its nature, transference is not the shadow of something that has been lived in the past. On the contrary, the subject as subjected to the analyst's desire, desires to cheat him about this subjection, making himself *loved by the analyst*" (Lacan, p. 229, emphasis added). Without the subject there is no experience of analysis. For analysis to take place it is necessary to accept transferential belief, named as *love*. The analyst learns in analysis that it leads us to become separated from narcissistic love, which is directed toward an image. Thus, what must be looked for is the love directed toward the subject. The desire of the analyst is one of the pieces of this enigma if one wants to succeed in managing transference. The task of the analyst is to maintain the Big Other, which we might read as *the culture*. From this point of view language is centralized. It ceases to be regarded as a means of "communication"; it becomes the very process of constitution of the subject. In a word, the subject or learner is constituted within a language community.

Applying transference to the teaching and learning of mathematics leads to the concept of *pedagogical transference* (Cabral, 1998). The pedagogical transference implies that listening is not unrestricted, as it is in the clinic, but is restricted to mathematical listening. In other words, the pedagogical transference is restricted to the possibility of attributing mathematical meaning to what is heard. Mathematical listening is defined as occurring when the listener is able to repeat the speaker's discourse until the speaker agrees that it is exactly what is meant. Of course, the speaker can always disagree and claim that something else was meant. Therefore the activation of restricted listening presupposes an agreement, tacitly established prior to the talking situation.

The pedagogical transference occurs when (1) the student manages to adjust the image of himself that he sees in the mirror, to his expectation of being loved by the teacher and (2) the teacher accepts this image as capable of being loved. This love is not to be trusted, since the student is only seeking the way to produce the right answer, so that the teacher and significant others will be satisfied, and that the student may gain recognition. The identification process of the subject with the image that he supposes to be loved by the teacher is called the *imaginary identification* and is denoted by i(a) where the "i" stands for "image" and the "a" for the object of desire around which the image is built. Schematically, in pedagogical transference, this identification is represented by the ability displayed by the teacher in producing answers and deciding what is right or wrong. It is an ability offered to the student as one to be imitated. In the pedagogical transference, the commitment is mingled with the ability to produce answers, leading generally to strongest learning efforts.

This implies, first, that, in dealing with affect and its connotations, language is absolutely primary, not a 'means of communication,' since the constitution of the subject (students, teachers, researchers) depends on it. Second, if we want our practice to have anything to do with affect, that is, if we want the student to be captured as an object of desire in the discourse of our practices, we need to stop talking and start listening to the student. Like Blanchard-Laville (1997, p. 168) the point is not to try to eliminate "non-sense." "Non-sense" is what prompts the moments of successive opening and closing of the unconscious, whose reality transference is expected to actualize. Let us see how this theory works through two examples of practice.

THE SETTING AND THE EXAMPLES

The basic unit of the psychoanalysis-inspired learning experience discussed is a weekly meeting assembling teachers and graduate students of a mathematics education program, undergraduate students of a mathematics teacher formation program, and mathematics teachers from the neighboring school district—in total from 10 to 15 people. These meetings are called *integrated sessions*. They started several years ago when an undergraduate student asked for help in solving his specific difficulties: this young man had failed all his freshmen courses. Gradually, other undergraduates joined in and took advantage of the teaching made available to him. Nowadays these meetings provide course credits for undergraduates as well as research material for graduate dissertations and papers. They provide the team with the experimental base to adapt Lacanian concepts for the classroom.

In these sessions, "teacher" and "student" are not labels attached to people, but positions of speech. Whoever is at the blackboard—generally an undergraduate student—is called "the student." The student is expected to work in order to produce some sort of knowledge, connecting a mathematical object to his or her lack of knowledge. "The teachers" are all those who put restricted mathematical listening into practice and provide guidance. The teachers' responsibility is to sustain the student's speech. These learning experiences presuppose that it is through speaking that one learns and through listening that one teaches (Cabral, 1998).

In the sessions the didactical, pedagogical and mathematical objects are treated simultaneously.

The First Case: Every Interior Point of a Subset of R is an Accumulation Point

I shall report on an integrated session in which a young man sought guidance with this problem from his mathematical analysis course. In the dialogue, the term "teacher" does not always refer to the same person. The teacher first oriented the student by asking for him to reproduce the definitions. He concluded that, if x belongs to int(X), then x belongs to the derived set of X, denoted by X'. When everything seemed to be over and done with, the student expressed the following doubt:

> **Student:** All right, I have proved that if this x is in int(X) then it is in X', but this is not enough, since here, in X', I have *all* the accumulation points and I proved it only for this particular x.

First, everybody interpreted that he had said "here in int(X) I have all interior points and I only proved it for this x" and thought that the difficulty hinged on the universal generalization law of logic: once we have proved for a generic x, the proposition is true for all x. But the student replied:

> **Student:** I understand that I have proved for this x which is any generic one, so that it is proved for all x. But there, in X', are *all* the accumulation points, not only this x.

What now? The discussion lasted for more than one hour. Everyone was eager to make a contribution. The students' and teachers' voices grew louder with enthusiasm. But at each turn the student said:

> **Student:** I know what you want me to say, but I am not convinced.

And he repeated his doubt. When the session time was over, one of the teachers asked permission to try out her approach. In a low voice, she asked the student to repeat the whole reasoning. He summarized it while the teacher wrote on a clean black board what the student said, always asking him: "Like this?" or "Is this right?" The teacher exhibited a genuine effort to understand every word the student said and did not write anything beyond what she heard. Finally the teacher expressed her doubt.

> **Teacher:** Where do I write this "all"? Here, at int(X) or there at X?
> **Student:** At X'.
> **Teacher:** This is the difficult point for us. If this "all" were here at int(X) we would know how to orient you. It is the old

story about proving that every cat has a tail. It does not suffice to pick one cat. You have either to bring them all or to consider that something is nothing more than a cat, a generic cat, and so on. OK?

The student agreed and the teacher went on, always in a low voice.

Teacher: But since you put the "all" here, at X', we don't really get you. Can you explain?

Her approach was by way of invitation to explain. The student thought for a while and looked embarrassed.

Student: I do not know what I want to say. [*The student confessed*]
Teacher: Try and say it. [*The teacher insisted*]

The essential point is that the statement "try and say it" was not uttered in a mood such as to mean: "now do you understand?" or "do you see your mistake now?" The meeting ended in a happy mismatch.

The Second Case: If I follow the Definition ...

In another integrated session another male student volunteered the following problem:
Let the sequence of functions $f_n:[0, \infty] \rightarrow R$ be defined by

$$f_n = \frac{x^n}{1 + x^n}.$$

Show that f_n converges point-wise and determine the limit function. Show that it does not converge uniformly.
The student showed proficiency in taking the limit for $0 \le x < 1$, $x = 1$, $x > 1$ and soon arrived at the limit function and its graph:

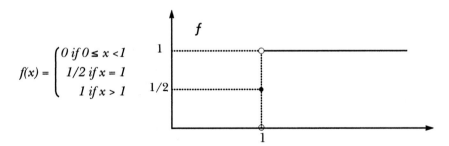

$$f(x) = \begin{cases} 0 \text{ if } 0 \le x < 1 \\ 1/2 \text{ if } x = 1 \\ 1 \text{ if } x > 1 \end{cases}$$

Figure 8.1.   The limit function.

He was asked to explain what he understood by point-wise convergence and uniform convergence. He wrote down the two definitions with some proficiency. He needed some help in order to explain that in uniform convergence N does not depend on x.

$$\forall \varepsilon > 0 \quad \forall x \in [0, +\infty] \quad \exists N \in \mathbf{N} \quad \forall n \geq N \quad |f_n(x) - f(x)| < \varepsilon$$
$$\forall \varepsilon > 0 \quad \exists N \in \mathbf{N} \quad \forall x \in [0, +\infty] \quad \forall n \geq N \quad |f_n(x) - f(x)| < \varepsilon$$

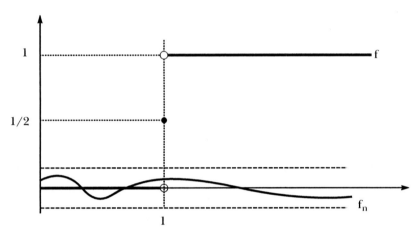

Figure 8.2.   The student's strip.

> **Teacher**: Well, then you have that for every x, the sequence $f_n(x)$
> converges to $f(x)$. But is the convergence uniform?
>
> **Student**: No, if it were uniform from this N on, all the f_n would
> have to stay inside this strip.

The teachers all expected the student to produce a picture like the one in Figure 8.3 and argue that then the functions f_n would fail to be continuous. The efforts to understand this student's image lasted for the duration of the meeting.

His justification was not helpful, because he simply insisted that his picture was what uniform convergence meant. He was asked to explain what he meant by "f_n being inside the strip." He pointed at $|f_n(x) - f(x)| < \varepsilon$ and said that this inequality meant that for every x, $f_n(x)$ should be between $f(x) - \varepsilon$ and $f(x) + \varepsilon$. We stressed the *every x* and asked him to produce his words on the picture. He seemed to stumble on the strip he had drawn for $x > 1$. We had to insist: "*choose one x; take ε of this size* (we drew a small segment), *mark $f(x) + \varepsilon$ and $f(x) - \varepsilon$, do it for another x*, etc. He finally erased his strip and drew the correct one around *f*. Next he was asked to show one f_n inside the strip. The request seemed to make no sense to him, as if no f_n could be inside this new strip. The session continued:

Suppose the convergence was uniform, then these f_n with n > N would be inside the strip. Draw one of them. This time he was asked to do it point by point for as many as eight or ten values of x, because he tried to avoid choosing x close to 1. Finally the graph of f_n inside the strip was drawn.

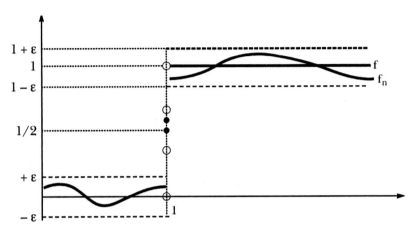

Figure 8.3.   Inside the strip.

It was hoped that he would recognize that the convergence could not be uniform because f_n would have to be discontinuous at 1.

> **Teacher:** Is the convergence uniform?
> **Student:** [*Looking at the picture*] No.
> **Teacher:** Why not?
> **Student:** Because then f_n would have to be inside this strip [*and he drew the strip of Figure 8.2 again*].

Those at the session endeavored to understand his thoughts. The following discussion ensued:

> **Teacher:** When you have one f and one ε, how many strips can you draw?

After some discussion he agreed that there was only one strip and that the strip for f would be the one in Figure 8.3. However this conclusion seemed to have put him in doubt that he was dealing with only one function: he started referring to f as "these functions."

> **Teacher:** You have to make up your mind. What do you see in this brace [of Figure 8.1]? Is it one function or are there three functions?

He decided that he would always "see" three functions there and named them g_1, g_2, and g_3. The teachers felt apprehensive about how to proceed. Fortunately, the problem statement stated *the limit function*, suggesting that there should be only one. The student then declared that the brace would always mean only one function to him. Next he muttered something about a unique value that $f(x)$ would have, if the convergence were uniform. This idea would partially explain his strip: he was seeing more than one value for f in it.

> **Teacher**: Are you saying that when the convergence is uniform the limit function must be constant?

He timidly nodded in agreement. He was then shown a clear example of a sequence converging uniformly to a non-constant function. He then looked at the graph and seemed annoyed by the discontinuity.

> **Student**: The strip cannot break.
> **Teacher**: Are you saying that if the convergence is uniform then the limit function cannot be discontinuous?

This time he agreed. The example $g_n(x) = f(x) + \frac{1}{n}$ taking f as the above limit function was offered. He was then asked to draw the graphs of these functions. He ended by agreeing that the convergence was uniform though the limit function was discontinuous.

> **Teacher**: Then, why is the convergence of f_n not uniform?

Once more he resorted to his original strip argument, but now he seemed less sure of it. He went back to the definition and, in his words, "opened" the absolute value: $f(x) - \varepsilon < fn(x) < f(x) + \varepsilon$. He seemed puzzled about how this inequality could hold and f_n could still be inside his strip.

> **Teacher**: This is what we are trying to understand. You say that there is only one function here [pointing at the graph], namely, this f. You say that with one G you have only one strip around this f. You look at the definition and say that if the convergence were uniform, $f_n(x)$ would be inside this strip. However when we ask if the convergence is indeed uniform, you say that f_n is not in this strip, but inside another one. Where does this new strip come from? Please tell us.
> **Student**: [*After some reflection*] Aha, now I see it [and he started explaining what he had realized]. If I follow the definition…
> **Teacher**: If you follow the definition… ? [Everybody laughed] What have you been following till now?
> **Student**: If I follow the definition the convergence is uniform.
> **Teacher**: Do you mean that if you follow the definition the f_n will be inside the strip and the convergence is uniform?

Student: [*Now emphatically*] Yes.

There was no hurry at this stage to correct the error. That correction was given later. Instead the following comment was made:

> **Teacher**: [*To everyone*] You see that mathematics is difficult because people do not abide by what they promise. They say, "here is a definition that I should use because it is part of a university course." When somebody asks me what uniform convergence is, it is this that I must answer. However, when I am alone by myself and I have to decide whether a certain convergence is uniform, then I have my own way to think about it. I may swear that I will follow the definition, as I may swear that I will obey the law, but Brazilians are just too smart to keep their pledges. We have a worldwide reputation of always been able to find a way around every difficulty.[1] In addition, law enforcement recalls the dictatorship time. We should rebel against such impositions, as obeying laws and following definitions. Mathematics educators should remove all impositions that make mathematics so difficult and hateful. Doing mathematics should be a light and pleasant activity, where everyone may say and write anything, without being reminded that this or that does not make sense or is wrong. Nothing should be "wrong." Human beings should be set totally free. Definitions may be there, we may swear to abide by them, but we know that they are intended only for Englishmen.[2] In practice, we do something else: we do it the Brazilian way.

BEYOND THE DATA

The theory elaborated by Lacan to explain the dialogic encounter with his patients may be characterized as the *dialectics of the subject and the other* (Lacan, 1973, pp. 205–239). This theory models what happens whenever a subject addresses an audience in a common language. Thus, in both experiences—analysis and learning—the phenomenon of language is brought to the fore. Contrary to common sense assumptions, in the proposed theoretical frame, it is not the speaking subject who decides the meaning of what has been said: it is not the speaker who produces the meaning. Granted, these ideas run up against common sense notions, where someone would naturally ask: Why did you not tell him? Common sense takes meaning as a one-sided production of the speaker. The idea is so strong that, for instance, some research reports only register the subjects' answers in cognitive experiments. The questions are not transcribed.

The reader might recall that in the first episode the student said clearly several times "here, in X', I have *all* the accumulation points" but we all answered as if we had 'heard' him saying "here in $int(X)$ I have all interior points." In addition to this, what the listener returns to the speaker is never exactly what the speaker wanted to hear. What the speaker hears as a response from the Other does not come back exactly as expected. Indeed, we expected the student to say "Aha!" An "agreement to agree" is necessary in order to overcome this communication difficulty, because the teacher does not actually have the exact *signifier* that was sent to form his response. In sum, the speaker becomes subjected to a flaw of the Other. This basic mismatch is constitutive of language: *meaning is a function of the Other* and a joint production of the subject and the Other.

The raw material for meaning production is what the subject says. Jokes, mispronunciations and "misconceptions" play a major role, especially when they are systematic and constitute *obstacles*. The second episode shows a clear instance of such an obstacle: the student had probably been exposed to a situation where the strip was drawn around a continuous function and retained this piece of certain valid knowledge so strongly that the joint production of a different meaning demanded a great effort from everybody.

These two cases were developed in integrated sessions that offered the possibility of commitments not available in traditional teaching. In the first episode the student seemed proud of his persistent question, proud of having 'found' such a doubt that nobody could dismiss. In the second episode the student seemed really in love with his continuous strip. Such persistent preferences transcend the symbolic domain; they are not translated into words and they constitute what is called *jouissance*. In the final remark of the second episode the teacher tries to point out to the student the nature of his difficulty, his *jouissance* organization rooted in the Brazilian culture. It is necessary to stress that such an intervention would not be tolerable in an analytic session; it would be classified as *savage interpretation*. However, since the integrated session is not a psychoanalytical session, we are still investigating its pedagogical value.

In the pedagogical transference of traditional teaching, the commitment is mingled with the ability to produce answers, leading generally to strong rote learning efforts. Lacan (1973) says that this mingling of ideological identification with the object of desire (*petit-a*) "is the safest structural definition of hypnosis that has ever been produced" (p. 245). By this Lacan means that in transference, the student is in a state of hypnotic trance, a most delicate situation. A student who produces the right answers according to mathematical canons may well represent the teacher's object of desire. Up to a certain point the teacher may pretend that the student's rote answers are produced according to mathematical principles and accept them. The transference situation becomes well installed. However, the teacher may be incapable of including among the representatives of her object of desire a "non-sensical" student-mathematics relation. The reader may experience difficulty with the student's objections such as "there, in X', are *all* the accumulation points, *not only this x*" in the first episode or, in the second, where the student produces a picture of uniform convergence in order to justify that the convergence is not uniform. Due to her inability to follow the non-sense, at some point, the teacher intervenes and the student awakes from the "trance" without having been prepared for this moment by an adequate analysis of the pedagogical transference. This is what happens when we "tell" the student his "mistake" as common sense suggests. It is not only the student's love message that is rejected; it is the student's unconscious commitments supporting this message—the image of himself is impaired. The super ego crashes and fear and anxiety emerge.

In the first episode, when the session was over, the ability of the teacher was praised, but she replied that she was really curious to see what the student was thinking. So, the teacher had assumed that there was indeed a meaning and she wanted to find it out. In capturing the problematic student-mathematics relation as an object of desire, the hypnotic situation became inverted. It was the teacher who was hypnotised by the student's discourse, trying to make sense of it to the satisfaction of the student. In this way the teacher distanced herself from the student's self image while taking the focus on the nonsense as her ideological commitment. Since the student could no longer remember the thinking behind the error, she confessed that she felt somewhat deceived. The student produced the "nonsense" himself. The "non-sense" was playing the role of her *petit-a.*

The point was made that in every previous attempt to reorient the student, the teachers had always presumed that they knew something about the student's mathematical difficulty (they assumed the position of the subject supposed-to-know) or intended to reach some foreseen point (they tried to rectify the transference). What I want to stress is that discourse is a *joint effort* of the teacher and the student in order to *sustain* actions and utterances (Baldino & Cabral, 1999).

I concede that the first student was mostly demanding some sort of affect, insisting on his initial relation of transference in spite of undergoing dozens of such sessions. From his previous experience with the team he realized that this was indeed possible. On being recognized in his attempt he received a form of love. However, he was not spared his suffering: no one told him that what he was saying was misguided, nor was he praised for abandoning his previous view. He had to take responsibility for his learning.

This treatment of the student might be viewed as either too lenient or too harsh. However, comments at the end of each session from the student audience were positive. Those comments tended to highlight Miller's (2001) argument: "Narcissistic love is that which is directed toward an image, while Lacanian love is that which is directed toward the subject" (p. 41). It is possible to interpret the situation as producing the effect of joining parts of a shattered body into one unit. The student has to experience, at least once, the vision of his entire body within the mathematical scene reflected on the Other. The situation resembles that of the child in front of the mirror when an adult, generally the mother, for the first time, points to the mirror and says "that one is you." While the adult thinks that she is saying "that one is YOU," Lacan stresses that the child will understand "that ONE is you" and will be able to collect all her life experience into one unit, the self.[3] "We do not know any other support to introduce the ONE into the world except the signifier as such, that is, as that which we learn to separate from its effects of meaning" (Lacan, 1975, p. 48).

It is important to stress that at any time the interpretation of student-teacher interaction was rooted in the description of specific mathematical problems in spite of the way they are presented in the dialogues. Another point to note is that the subjects' feelings were not described from a point of view of a superior conscious ego that in most cases is identified with the teacher's position. In many analyses of classroom episodes the teacher assumes an external and neutral position with respect to the processes of learning mathematics. This was clearly not the case described here.

CONCLUSION

In this chapter I have looked at the relationship between affect and cognition. In going against the discipline's customary approach to the affective domain defined by a range of beliefs, feelings and moods, I drew on psychoanalytic theory to introduce a different perspective of learning. Using the notion of transference—one of Lacan's four fundamental concepts of psychoanalysis—I developed the concept of *pedagogical transference* to capture what takes place between the teacher and the student in the learning process that includes a mathematical object. Two examples from my own work illustrate the psychoanalysis-inspired approach.

I consider the psychoanalytic method provides a sound explanation of the learning process. For one thing, the approach is able to take into account both the conscious and unconscious aspects of learning. In particular, learning, in the psychoanalytic formulation, is not founded on an autonomous conscious learner in subordination to a superior conscious ego, identified with the teacher's position. For another, pedagogical transference, as construed by mathematics education, overlooks important aspects between the parties in the learning process. In postmodern times like these it is my suggestion that the discipline look toward psychoanalytic theory, not only for explanations about learning, but for understanding many of its other processes.

NOTES

1. "Dar um jeitinho" as we used to say.
2. "Para ingles ver," an idiomatic expression of the 19th century referring to laws against slave traffic imposed by England but that were not enforced by Brazilian authorities.
3. This function of the ONE is stressed by Lacan (1975) in dealing with the question "what is a signifier?" which he rephrases "what is ONE signifier?" The function of the ONE as a totalizer of collections is also developed by Hegel (1929, pp. 177–182). It lies at the root of the distinction between discrete and continuous quantity and pervades all mathematical learning.

REFERENCES

Baldino, R.R & Cabral, T.C.B. (1999). Lacan's four discourses and mathematics education. In O. Zaslavsky (Ed.), *Proceedings of the 23rd Conference of the International Group for Psychology of Mathematics Education* (Vol. 2, pp. 57–64). Haifa: Technion Institute.

Blanchard-Laville, C. (1997). L'Enseignant et la Transmission dans l'Espace Psychique de la Classse. *Recherches en Didactique des Mathématiques, 17*(3), 151–176.

Breen, C. (2000). Becoming more aware: Psychoanalytic insights concerning fear and relationship in the mathematics classroom. In T. Nakamara & M. Koyama (Eds.), *Proceedings of the 24th Conference of the International Group for Psychology of Mathematics Education* (Vol. 2, pp. 105–112). Hiroshima: Hiroshima University.

Cabral, T.C.B. (1998). *Contribuições da Psicanálise à Educação Matemática (Contributions of Psychoanalysis to Mathematics Education)*. Doctoral Thesis. São Paulo (BR): USP.

Cabral, T.C.B., & Baldino, R.R. (2002). Lacanian psychoanalysis and pedagogical transfer: Affect and cognition. In A.D. Cockburn & E. Nardi (Eds.), *Proceedings of the 26th Conference of the International Group for the Psychology of Mathematics Education* (Vol. 2, pp. 169–176). Norwich (UK): The University of East Anglia.

Da Rocha Falcão, J.T., & Hazin, I. (2001). Self-esteem and performance in school mathematics: A contribution to the debate about the relationship between cognition and affect. In M. van den Huevel-Panhuizen (Ed.), *Proceedings of the 25th Conference of the International Group for the Psychology of Mathematics Education* (Vol. 3, pp. 121–128). Utrecht: Freudenthal Institute.

Lacan, J. (1973). *Le Séminaire de Jacques Lacan. Livre XI. Les quatre concepts fondamenteaux de la psychanalyse. 1964.* Paris: Editions du Seuil.

Lacan, J. (1975). *Le Séminaire de Jacques Lacan. Livre XX. Encore.* 1972–1973. Paris: Éditions du Seuil.

Lacan, J. (1977). *The four fundamental concepts of psycho-analysis.* London : The Hogarth Press.

McLeod, D. (1992). Research on affect in mathematics education: A reconceptualization. In D. A. Grouws (Ed.), *Handbook of research on mathematics teaching and learning* (pp. 575–596). Toronto: Macmillan Publishing Company.

Miller, J-A. (2001). *Pure psychoanalysis and applied psychoanalysis.* NYC: Lacanian Ink 20, 4–43.

Schlöglmann, W. (2002). Affect and mathematics learning. In A.D. Cockburn & E. Nardi (Eds.), *Proceedings of the 26th Conference of the International Group for the Psychology of Mathematics Education* (Vol. 4, pp. 185–192). Norwich (UK): The University of East Anglia.

Vinner, S. (1997). From intuition to inhibition: Mathematics education and other endangered species. In E. Pehkonen (Ed.), *Proceedings of the 21th Conference of the International Group for Psychology of Mathematics Education* (Vol. 1, pp. 63–78). Lahti, Finland.

Vinner, S. (1996). Some psychological aspects of professional lives of secondary mathematics teachers: The humiliation, the frustration, the hopes. In L. Puig & A. Gutiérrez (Eds.), *Proceedings of the 20th Conference of the International Group for the Psychology of Mathematics Education* (Vol. 4, pp. 403–500). Valencia: University of Valencia.

Wilson, D. (1995). The transference relation in teaching. *Chreods 8*, Retrieved July 17, 2002, from http://s13a.math.aca.mmu.ac.uk/chreods/Issue_8/Dave_W/Dave_W.html.

part III

POSTMODERNISM WITHIN THE STRUCTURES OF MATHEMATICS EDUCATION

Part III relates to the postmodern analysis of the limits of established ways of thinking about reality and truth. The four chapters in Part III are concerned with processes, structures and visions of mathematics educational practice. The focus is, in turn, on representation, on the relationship between power and knowledge, and on how that relationship intersects with notions of identity.

CHAPTER 9

IDENTIFYING WITH MATHEMATICS IN INITIAL TEACHER TRAINING

Tony Brown, Liz Jones, and Tamara Bibby

ABSTRACT

This chapter addresses issues of identity among trainee teachers as they progress through university into their first year of teaching mathematics in primary schools. We examine teacher identity and suggest that it is produced at the intersection of the trainees personal aspirations of what it is to be a teacher and the external demands encountered en route to formal accreditation. We also suggest that participation in a teaching institution results in the production of discourses that serve to conceal difficulties encountered in reconciling those demands.

INTRODUCTION

Int: So what do you see the purpose of teaching as being?
Nathan: As being just educate children and make a difference in their lives really—just to make things—it's just a very small part of their lives you know the year in which I

Mathematics Education Within the Postmodern, pages 161–179
Copyright © 2004 by Information Age Publishing
All rights of reproduction in any form reserved.

teach them, really but hopefully to make a difference
really—you know I can see a purpose in it.

Int: What sort of a difference would you want to make?

Nathan: To give them, I don't know, confidence in themselves to
enjoy school, enjoy learning, enjoy books and you know
just have a real … just have an enthusiasm for life you
know not be resigned to thinking things are not worth
doing you know and yeah preparing them for secondary
school so they don't go having a negative attitude toward
school.

This chapter addresses issues of identity among trainee teachers as they
progress through university into their first year of teaching mathematics in
primary schools. We examine how we might conceive of the trainees confronting mathematics in the context of government policy instruments. We
suggest that teacher identity is produced at the intersection of the trainee's
personal aspirations of what it is to be a teacher and the external demands
they encounter en route to formal accreditation. We also suggest that participation in the institutions of teaching results in the production of discourses that serve to conceal difficulties encountered in reconciling these
demands with each other.

We commence with a brief account of an empirical study involving
trainee primary school teachers. This is followed by a look at how student
teachers experience external demands as constraints on their personal
aspirations. We, however, urge caution in inspecting the trainee's accounts
of this experience, by suggesting that the accounts mask anxieties emerging in a difficult transition. We then offer the theoretical notion of "identity" guiding our work (cf. Brown & Jones, 2001). Readings based on
interview transcripts relating to student conceptions of mathematics are
then offered. Here our task is to show how "identity" becomes a means for
reconciling the past with the present and the future. Our suggestion is that
because certain reconciliations have been effected, transitions between the
two sites of learner and teacher, are made possible. Finally, we consider
how mathematics, shaped as it is between multiple sites, emerges within
the teachers' appropriation of the social discourses surrounding their
teaching of the subject. We also consider here how teacher professional
development might be better understood.

THE EMPIRICAL STUDY

We draw on two studies undertaken within the B.Ed. (Primary) program at
the Manchester Metropolitan University in the United Kingdom (UK).
The empirical material provided a cumulative account of student transi-

tion from the first year of training to the end of the first year of teaching. The specific interest in the discussion which follows is on how the students'/teachers' conceptions of school mathematics and its teaching are derived. In particular, we explore the impact that government policy initiatives relating to mathematics and Initial Teacher Training (ITT), as manifest in college and school practices, have on the construction of the identities of the primary student and first year teacher.

The first study spanned one academic year (Brown, McNamara, Hanley, & Jones, 1999). We interviewed seven/eight students from each year of a four-year initial teacher education course out of a total annual cohort of some 200 students. Each student was interviewed three times at strategic points during the academic year: at the beginning of the year, while on school experience, and at the end of the year. The study took the form of a collaborative inquiry between researcher and student/teacher generating narrative accounts within the evolving students'/teachers' understandings of mathematics and pedagogy in the context of their past, present and future lives. The second study, which followed a similar format, spanned two academic years. In the first year of the study a group of 4th year student participants ($n = 37$) was identified. Each student was interviewed three times during this year. The group included seven students involved in the earlier project, five of whom were tracked for a total of four years. In the second year of the study a small number of these students ($n = 11$) were tracked into their first teaching appointment. Each of these students was interviewed on two further occasions. These interviews monitored how aspects of their induction to the profession through initial training manifested itself in their practice as new teachers.

Specifically, students involved in the research were those who were training to be primary teachers and who, as part of their professional brief, would have to teach mathematics. Significantly, while all the students who were interviewed held a General Certificate of Secondary Education (GCSE) (16+) mathematics qualification as required for entry to college, none had pursued mathematics beyond this. Nor had any of the students elected to study mathematics as either a first or second subject as part of their university course. The research set out to investigate the ways in which such non-specialist students conceptualize mathematics and its teaching and how their views evolve as they progress through an initial course.

RECONCILING PERSONAL ASPIRATIONS
WITH EXTERNAL DEMANDS

Nathan, whom we met at the beginning of this chapter, is a primary teacher. How might his personal aspirations be understood? In that quote, he is talking about the sorts of motivation that underpin his developing practice as a professional teacher. His comments point to a desire to participate in an educational enterprise aimed at making things better for the learners in his class. He hopes that they will "enjoy school" and as a result builds an "enthusiasm for life." Such sentiments seem unsurprising for someone entering the profession. The motivation of buying into such a strong mission must be very appealing to those mapping out a future career. In this perspective the task of teachers is not just about raising standards according to the latest government directive—the motivation is to improve the quality of educational experience more generally and hence the subsequent lives of the pupils. Teaching is about empowering young learners and as such can be seen as a very worthy profession, around which one can harness more personal aspirations such as feeling one has social worth and a clear identifiable professional purpose.

Nevertheless, Nathan is obliged to work within a professional framework. Within a broader perspective of social improvement, the role of individual teachers often takes second place to the wider social agenda. Such individual teachers become participants in a collective program where their personal aspirations need to be filtered through a set of socially defined demands. Such demands get to be meshed with the requirements for accreditation as a teacher and the regulations governing everyday practice as a teacher in schools. Trainee teachers in the study seemed less enthusiastic than Nathan, when it came to having their individual practices as teachers and mathematicians gauged against the externally defined definitions of what it is to be a teacher, as for example, in government sponsored inspections carried out by the UK Office for Standards in Education (OfSTED). We offer a few brief comments from the study to illustrate how the new teachers perceive the potential conflict between personal aspiration and external demand.

> It feels as if they're checking up on you all the time, yeah, they're not leaving it to your own professionalism … but the university have to cover their own backs don't they, with OfSTED [inspectors] coming.

> But I am here for the children, OK I am to meet the criteria, but I am not here to prove to the OfSTED that I can do maths.

The study coincided with the introduction of the National Numeracy Strategy (DfEE, 1999), a high profile government initiative defining the content and conduct of mathematics lessons in great detail. While most students regarded the Strategy and its daily "numeracy hour" as "very useful," it resulted in nearly all schools and individual teachers in the sample abandoning their own more personalized schemes of work. Changes to teachers' work impact on learners. Hardy (this volume) gives a clear expression of how normalizing and surveillance practices associated with the UK reform play out in the classroom. In our study it was not uncommon for some teachers to find the Strategy somewhat over-prescriptive:

> The numeracy hour, it's so prescriptive as to what you have to do, when you have to do it and how long you do it for, so it shapes the whole numeracy hour of every day of every week of school year.

> We don't always stick to it exactly because I feel it's a little bit too restricted.

But for many trainees interviewed the idyll of teaching encapsulated by Nathan was not compromised by recent government requirements alone. The idyll was also somewhat punctured by a sometimes unwelcome, yet long-standing component of the overall job description of a primary school teacher, namely, the actual need to teach mathematics in the first place. Many had significant emotional turmoil in their own experience of mathematics while pupils at school, where it seemed that some had received excellent training for becoming compliant individuals:

> It was just a case of doing the sums but you didn't realise why you were doing the sums. I think the teacher's role played a big part in it as well because the atmosphere she created, it wasn't a very, it was just a case of if you can't do it, you should be able to do it now. It wasn't very helpful or you didn't feel like, she wasn't very approachable, you didn't feel like you could go to her and say I'm having trouble with this and I need some help, it was just a case of don't even bother going to a teacher, just very much a case of you have to meet the standard and if you don't then you're a failure. So I didn't really enjoy maths at all.

Attitudes such as those expressed here were very common in the study. Those expressions worked against any ease about producing a conceptualization of teaching through which personal aspirations could be achieved.

WHAT DO THE DATA TELL US AND WHAT MIGHT THEY CONCEAL?

In analyzing such data, however, there seemed to be a need for us as researchers to adopt a certain amount of caution (cf. Convery, 1999). We felt reluctant to accept all of the accounts provided at face value. We wondered what could be concealed in such stories? Story telling can be used as a support device to sustain teacher learning (e.g., O'Connell Rust, 1999). But surely in the last data extract the interviewee did not have just one teacher, introduced here as "she." The trainee appears to be personifying his entire experience of many teachers in just one teacher who is required to carry the weight of this individual's perceived suffering at school. We may wonder which narrative devices individuals employ when they are requested to recount experiences that happened some ten to twenty years earlier. For what reasons do they construct such images of themselves and what present demands are concealed in these images? How do teachers tell the story of their lives to rationalize their current motivations—their futures? Freud might suggest that a repetition of such a story may be a form of resistance, an insertion of a fixed image, which blocks off the possibility of building memories in a more creative way (cf. Ricoeur, 1981, p. 249). The reworking of memory into a story is not the memory of a linear narrative "as it was" but rather a probing that creates something new; a present day building of the past, shaped by current motives, but perhaps also distorted by things the student would rather not confront. Consequently, our analysis adopts a cautious attitude to the data presented in seeking to build a better picture of how student teachers account for their own transition into professional teaching.

All transitions are potentially problematic. But the movement between the two locations of learner and teacher seemed especially fraught for students entering initial training as primary teachers, who so often perceive mathematics as a source of anxiety. This chapter marks an attempt at having some appreciation of the various kinds of "selfwork" (Stronach & MacLure, 1997, p. 135) that are undertaken by students when making the move from being a 'learner of mathematics' to becoming a "teacher of mathematics." Shortly, we shall suggest that for such students "identity" can be seen as a key feature in easing those tensions that lie between the two sites. Subsequently we try to show how particular constructions of the "self" are used to surmount and negotiate hurdles and boundaries. This includes a perceived lack of mathematical competence. Our main interest then is on perceptions of the "self" and how this, in relation to mathematics, is talked about, described and generally theorized. In part, this involves looking at the kinds of emotional baggage that center on mathematics and which students have collected over a period of time. We are particularly interested in those ways "identity" seems to assist in both accruing and jettisoning this baggage. We also note and discuss how accounts concerning past mathematical experiences are filtered through current perceptual frameworks. The crisscrossing between present perceptions of mathematics and the self, memories of mathematics and the self, and how together these feed into and help fashion future constructions of the self as "teacher of mathematics," is what we attempt to address.

In exploring the uneven territory between being a "learner of mathematics" and a "teacher of mathematics," we offer a series of interpretations based on interview transcripts. These interpretations are personal readings and cannot be considered as "hard edged analyses" or indeed "authoritative accounts." Despite this, however, we believe that these readings make tentative steps toward increasing our understandings about "transitions" and "identity." In offering examples of how a teacher makes sense of interactions in her own classroom we hope to illustrate an approach that other teachers might pursue in unfolding the discursive layers present in their own practice. First, however, we shall now say a little more about how we are using the word "identity."

IDENTITY

Identity should not be seen as a stable entity—something that people *have*—but as something that they *use*, to justify, explain and make sense of themselves in relation to other people, and to the contexts in which they operate. In other words, identity is a form of argument. As such, it is both practical and theoretical. It is also inescapably moral: identity claims are inevitably bound up with justifications of conduct and belief (MacLure, 1993a, p. 287, author's own emphasis).

The notion that "identity" is something people *use*, as outlined by MacLure, became a significant research theme. Briefly, perceptions of what constitutes "identity" are seen as "constantly being produced anew within different and competing discourses ... more fluid and drifting than had previously been assumed by reproduction theorists" (Haugh, 1987, p. 17; see also Davies, 1989; Lather, 1991; Stronach & MacLure, 1997). For these writers, questions to do with "true" or "real identities" are "unaskable." Rather, the perspective held is that "identity" is a "shifting and erratic performance" (Stronach & MacLure, 1997, p. 58).

Our use of the notion was also guided by the work of Zizek (1989) and Laclau and Mouffe (2001) who see "identity" as an always unsuccessful attempt to weave together specific identifications with various social demands. Demands for trainees in the UK working on primary mathematics are complex. They include, for example, meeting school requirements, meeting university requirements, being popular with children, pleasing parents, building an enjoyable conception of mathematics, performing adequately on the Numeracy Skills Test for teachers, achieving personal aspirations, following the National Numeracy Strategy adequately, getting through the Office for Standards in Education (OfSTED) inspections, minimizing teacher-pupil anxieties relating to mathematics, and, not least, teaching up to nine other curriculum subjects. Teacher identity then might be seen as the outcome of trying to reconcile these complex demands. The teacher might *use* a particular account of this reconciliation according to the demands of a specific domain. These might involve, for example, meeting college requirements or defending a lesson to a school inspector. Such an approach is quite different to more traditional conceptions within mathematics education research in which psychologically unitary teachers encounter groups of individual learners. The conception does however echo the work of Walkerdine (e.g., 1988) and some of the more socially oriented work to have appeared since (e.g., Walshaw, 2001).

In our analysis, those ways that the "self" perceived the world, including certain worries concerned with the learning and teaching of mathematics, became, in our view, central to how such concerns were confronted and addressed (Munby, 1986; Schon, 1979). Taking note of the figurative language that was used by students when talking about themselves, particularly in relation to mathematics, allowed us to glimpse some of their beliefs and orientations about learning and teaching. After all, mathematics, as such, does not exist in any tangible sense but nevertheless produces tangible effects as though it does exist. Mathematics does not impact on our lives as mathematics *per se* but rather through the social practices that take up mathematics into their forms (cf. Brown, 2001). Such social practices cannot be separated from personal engagements in them and the affective products of such engagements. Mathematics itself is thus necessarily shaped through the often emotionally charged activity that gives it a form (Bibby, 2002). As an example, the observed trainee teachers often presented a fairly clipped "didactic" version of mathematics, anxious as they often were about opening it up as a field of more creative inquiry. In some respects, this research echoes that of Kagan (1990) who, in seeking to develop alternative ways to evaluate newly qualified teachers' thinking, focused on their choice of metaphors. These were perceived as reflecting how they characterized "the nature of learning, the teacher's role in the classroom and the goals of education" (Kagan, 1990, p. 423). In this way, beginning teachers' metaphors gave some insights into how they had filtered and modified their university training.

The idea of identity as being *used* as a "form of argument" (MacLure, 1993b), we felt, could assist us in negotiating the boundary between student and learner (Jones, Brown, Harley, & McNamara, 2000). As a consequence, particular attention was paid to those parts of the texts in which the students talked about themselves as "learners of mathematics" and where they foresaw themselves being "teachers of mathematics." The methodological strategy used when analyzing the data is loosely derived from both conversational analysis and ethnomethodology. Both these approaches, while having certain distinctive characteristics, nevertheless share the view that language, action and knowledge are inseparable (Stubbs, 1983, p. 1). Our studies were not undertaken to find the "true" identities of the students, nor were they undertaken to find the "truth" about transition. Rather, our efforts were directed at unearthing those ways notions of self get talked about and how such notions become the means for negotiating and staking out particular claims, and become "theorized in discourse" (MacLure, 1993b, p. 377).

As we have seen, identity is also necessarily social. The argument of one's identity is entwined with an assertion of how one fits in, or does not, with one's perceived community. As Scheff (1994) contends, human consciousness is social "in that we spend much of our lives living in the mind of others without realizing it ... We always imagine, and in imagining share, the judgments of the other mind" (p. 45). Giddens (1991) draws on Scheff in discussing the notion of shame—the signal of a broken social bond. Shame, he contends, highlights the problems in the story we tell as our identity claim and the attendant feelings of "personal insufficiency" and the experience of loss of "integrity of the self" (p. 65). As Giddens (1991) puts it: "Shame directly corrodes a sense of security in both self and surrounding milieu" (p. 153). This work by Scheff and by Giddens has been discussed in relation to primary teachers of mathematics by Bibby (2002). We may ask, however, how such social bonds are defined? Can we see trainee teachers as taking control of their own personal and professional development and hence of their mode of participation in the community? Maybe to some extent, although, in the UK at least, so much of the teacher's task is externally defined by government policy apparatus. Participating in the maintenance of a social bond can so often become a task of compliance. As such, arguments of identity can become claims that we are meeting the expectations of others. We face shame if we do not.

Two key foci emerged from our readings of the transcripts; first, the students use of past, present and future events when accounting for themselves, and second, in describing their past mathematical experiences, negative perceptions of self were resituated as positive traits. We now go on to suggest that, by displacing certain negative perceptions and locating them as a positive term, transitions between the two sites of learner and teacher can be made possible.

TAMING MATHEMATICS

Question: How do you feel about Mathematics?
Answer: Maths is the demon that jumped out of the closet and licked me in the face. (3rd year student)

Demons are abhorrent creatures. They instill fear and are best avoided. Yet mathematics as a demon has managed to "lick" this student. Does this imply that the demon has been tamed and that some kind of affection lies between the student and the subject? Has the student's own fear of the subject been licked, and, if it has, how were the transitions between fear and friendliness, abhorrence and affection made?

The four transcripts that feature in this section were chosen because the students themselves share certain similarities and the transcripts reflect concerns with learning and teaching mathematics (cf. Jones et al., 2000). All four students, one from each year of the course, were women who, when starting the course, were aged between 18 and 19. Three of them had gained a minimum grade pass in mathematics (GCSE) at school while the fourth had acquired a slightly higher grade. All four had expressed a dislike for mathematics when they were at school and each of the students maintained that they lacked competence in the subject.

The interviews began with this question: What is the first thing that comes into your mind when you think of maths?

> I know I'm not very good at it … It's a way of adding and multiplying and taking away certain things … maths relates to numbers … it's so big. (Yr. 1)

> Maths is scary … I've always not been wonderful at maths. (Yr. 2)

The second year student then expanded on "what maths is":

> … numbers, problems, day to day activities. I know maths is involved in more or less everything I do in my life. We talked about that in lectures … there's the very complicated side of maths … when you're sat down and doing sums intensely, GCSE, algebra, that sort of thing … but if you're going back to the roots back to the simplest basic points of maths then it's to do with day to day problems and helping you through life … we talked about that … it's sorting out of things as well as the complicated side of things … organizational skills … sorting washing, blacks and whites … that's all maths … common factors or differences … that's maths … I didn't realize this before I came on the course … we've unpicked a lot on the course … and it's made me think maths isn't just scary numbers on a piece of paper which I used to think. (Yr. 2)

This student appeared to have developed a means of managing mathematics and, in part, her strategy was a consequence of college sessions. Mathematics, it would seem, was conceptualized as a series of binary oppositions. On the one hand there was the "very complicated side of mathematics" while on the other, there was the "simplest basic points." Using these two polarizations the students' responses could be presented as follows:

Complicated maths	Simplest basic points
sat down	active
doing sums	going back to roots
algebra	day to day problems
GCSE	helping you through life

pieces of paper	common factors
scary numbers	differences in things
	numbers, money
	grouping of things
	sorting out of things
	organization

It would seem, that through a process which is captured in the statement: "We've unpicked a lot of things on this course," the student made certain moves. This was a collective move in which she and her year group, together with the college tutor, worked together at "unpicking" mathematics so that aspects of it may be valorized. By way of the discursive practices of college mathematics, the student was motivated to leave behind that "scary maths" which is located in and associated with "doing sums intensely." Rather, she found herself moving backwards to the "roots" and to the "simplest basic points" in order that she might progress forwards toward teacherhood. And, as she traveled, there is, we believe, a sense of her beginning to collect some of the cultural baggage which has come to be associated with primary mathematics: certain terms, for example, "groups," "common factors," "organization" signal her entry into the discourse of primary mathematics. In effect, emotional and cognitive shifts are taking place within the self. There are the internal realizations that mathematics both exists and (importantly) can be understood even within mundane activities of everyday life. Simultaneously, external changes also occur; she is now beginning to sound like a primary teacher.

And what of the other students? How did they use notions of "identity?" It appeared that the fourth year student also dichotomized mathematics. She polarized the subject as either "do able" or "not do able" and by implication, subdivided mathematics into that which can/cannot be understood. She said:

> Simple calculations ... adding, subtraction, multiplication and division I can do, no problem. When you get into algebra ... I can't do it ... it's the more complicated things like statistics that frightens me ... I love addition because it's simple. I do things like area, capacity and volume because they're practical. I liked trigonometry ... you were given a question—you had a triangle in front of you and you could see that one of the sides or one of the bases was going to be longer than the other or whatever so you could work out roughly what it was going to be, whether it was going to be a reasonable answer or completely out of this world ... whereas algebra ... it doesn't really mean anything.

Her response could be arranged as the associations between:

Not do able maths	**Do able maths**
complicated algebra	simple basic calculations
fear	love
completely out of this world	reasonable answer
doesn't mean anything	means something

Meanwhile, the third year student posited the following theory that, so it seemed, helped in explaining her lack of mathematical competence:

> Somewhere along the line I just think that I've not understood it properly ... I personally feel that maths—to know how to do things you have to understand it in you as a person ... Sometimes I ponder over it and then I think I should know this anyway.

What is being implied here? Does the student, for instance, conceptualize the learning and understanding of mathematics as occurring along a linear developmental line? So that when she does master a particular problem her success is never read in fulsome terms. Rather, she thinks: "I should know this anyway." That is, she should have learned "it" at some specific or particular point *en route* to the present. She should already have been the person she now perceives herself to be. By constraining herself within a particular way of perceiving mathematical knowledge and its development, it would seem that a lack or gap will always exist between herself and the idealized mathematics student "who can understand it in you as a person."

Echoes of this notion of "understanding it in you as a person" could be found in the transcripts of the other students. There were, of course, variations in the ways that this was expressed. For example, the first year student categorized people as either "mathematical sorts of persons" or "arty sorts." Furthermore, because she defined herself as an "arty sort" she considered that this curtailed her chances of fully understanding mathematics. To quote:

> I think if you sat there and learnt and learnt and learnt I still don't think you could change the way you were. I don't think you can suddenly become a mathematical sort of person. I mean, I had tutoring for my GCSE and I had a lot of help from my teacher and no matter how much they explained things it still took me a long time ... other people got it just like that.

Similarly, the second year student talks about her brother as being able to do mathematics "just like that." He, it seems, "doesn't spend hours doing maths, but when he has to do it, it comes, just like that."

What are some of the consequences of these perceptions? What, for example, are the effects of placing oneself in the "not-capable" category?

One reverberation, which is highlighted in the third year transcript, is that mathematical achievement is perceived as paradoxical; success is always shadowed by failure: "she should have known it anyway." Similarly, the fourth year student found the learning of mathematics "very, very hard. For some people it just *naturally* clicks but I have to work and work and work at it" (our emphasis).

What are the implications for being a "teacher of mathematics" when the student has located herself within the "not-capable of mathematics" category? To us it seems that, rather than being perceived as a hindrance, this particular construction might be a strong motivating force. Thus, the first year student, who it should be remembered has had no college input, foresees that because she has "struggled so much, I think it would benefit me."

She then goes on to map out certain ideas for the teaching of mathematics:

I'd want to give them as much of the basics as I could because I think that would prepare them more ... I'd do it very practically. I'd say count these and count these ... what happens when you put them together? I wouldn't say "now add them up". I'd say: "what happens when we move this pile of bricks to this pile?" How many are there altogether?

The third year student, on the other hand, would "Try to understand where the children were coming from and where they got their ideas from to start with ... and I would ... break things down step by step rather than everything just seeming like taught as a whole." The fourth year student attributes an appreciation of the learner's perspective to helping her make the transition from "learner" to "teacher":

And it was kind of ... we are going to teach you how to learn this the same way that children will and that gives you a very good understanding of how the children learn maths as well ... you've been through that same process as you are going to teach the children and you know what to expect and you know broad outcomes of what might happen.

From this, we offer the following speculative thoughts: it would appear that for the fourth year student, shifts in locations including that between being "learner of mathematics" and being "teacher of mathematics" are not linear processes. In order to take on the future role as a teacher, the student implies that within the context of college mathematics sessions, she is repositioned as "the child." Such repositioning is premised on the assumption that as "the child" she can then attend to the "basic" and the "simplest points of maths" and in so doing she can leave behind all those negative aspects of mathematics. Upon entering school as a teacher, it is assumed that she will be able to demonstrate that she is indeed a teacher.

She will, for example, be able to control, organize and structure the primary classroom (see Brown et al., 1999, p. 313). But when it comes to the teaching of mathematics, internally she will be in many ways "the child" and it is this imaging of herself that will provide her with the confidence to teach.

Our rereading of the scripts is interpreted in this way: because of their own struggle with mathematics, the students were therefore determined to deliver the subject in terms other than their own experiences. As teachers, they will learn from their own gaps, omissions and lack in the subject. In effect, they will take all those things that in the past they have perceived as preventing them from developing mathematical competence and they will assert the authority of the "opposition." So the students did not, it seemed, want to become "mathematical" types. Rather, they appeared to draw strength from being the Other to this construct (Walkerdine, 1990, p. 62). And it would appear college helps to strengthen this persona:

> I didn't enjoy it because it was ... complicated ... intense, difficult, hard, didn't like it, boring ... so I thought from that well ... the children I was going to be teaching, I don't want them to be taught like that so I've been thinking about different ways of teaching which has come from University, they've helped in say in practical sessions, relating it to the home ... you use maths every day in everything that I never thought of ... washing, sorting out, organizational skills, variation in things, differences in things, common factors in things like three people have got brown hair, that's maths, it's relating it to just people. It doesn't have to be difficult like I did at GCSE to be able to understand maths, so I thought I like this approach, I enjoy it, it's easy to relate to, it's not tedious, it's interesting ... went on my school experience ... did a practical approach and it worked so therefore I've got confidence, I know what works, I know I have to go into everything thoroughly before I teach but as long as I make it interesting, don't let the children lose it, get bored, then it should be O.K. (Yr. 2)

Our readings in this section, we believe, can work at focusing attention on the significance of mathematics teaching identifications. For the students we have met, in order to succeed, mathematics must feature "in the genes" (be part of your identity, make-up). It either just "clicks" or it doesn't. If you are an "arty person," any current success in mathematics tends to be shadowed by the failures of the past and in this way, future experiences with mathematics are always prescribed. We would like to suggest that the nonspecialist trainee teacher, destined to include mathematics in her professional repertoire, appears to be wedded to the failed pupil, but seeks to revoke those characteristics of mathematics classrooms that are associated with failure. In some measure, this means declining to assume the identity of "a mathematical sort of person" frequently pathologized in the figures of the mathematics teachers assembled from the past. First, mathematics as a demon is powerful and through various ways it subjugates the student and fills that student with fear and loathing. But, in jumping out of the closet, mathematics is "outed." It is removed from the dark and abstract underworld and in the light it is possible to see the softer side of mathematics. This aspect of mathematics, besides being fun, is also basic and practical. In fact, mathematics is so friendly, besides letting it loose with children, you can, if you are so inclined, let it lick you.

MOVING TO A SOCIALIZED MATHEMATICAL IDENTITY

More broadly, within the UK, the use of mathematics curriculum materials has become high profile and rigorously encouraged. There are many accounts of mathematics, ranging from those built within the discourse of such government-sponsored materials to others generated more by the trainees themselves. Meanwhile, training institutions, schools, mathematicians, employers and parents all have some say in what constitutes school mathematics. For the trainee teachers interviewed in the studies, it seems impossible to appreciate fully and then reconcile all of the alternative discourses acting through them. In confronting the disparity between these alternatives, we found that the trainees produced an image of themselves as functioning professionals, in which the failure to reconcile perspectives was swept under the carpet. The individual trainee may, for example, buy into official story lines and see their "own" actions in those terms. We acknowledge that the interviewing process may have contributed to these responses, although we would prefer to think we had taken all reasonable measures to avoid collusion. By buying into official lines, the trainees may subscribe to intellectual package deals laid on for them rather than see the development of their own professional practice in terms of further intellectual and emotional work to do with resolving the contradictory messages encountered. As one teacher commented in carrying out research for a higher degree: "Why do we need to do research to find out what good teaching is when the government is telling us what it is?"

It seems, then, that any supposed resolution of the conflicting demands cannot be achieved without some compromises. Certain desires will always be left out. The teacher may nevertheless feel obliged to attempt such a reconciliation and to have some account of her success or otherwise. As an example, for so many of the trainees interviewed, mathematics was a subject that filled them with horror in their own schooling. Yet such anxieties seemed less pervasive once the trainee had reached "Qualified Teacher Status." How has this been achieved? It would seem, as in the last section, that those who so often had ambivalence toward the subject of mathematics do not continue to present themselves as mathematical failures. Rather, they tell a story in which their perceived qualities have a positive role to play. For example, "I like to give as much support as possible in maths because I found it hard, I try to give the tasks and we have different groups and I try to make sure each group has activities which are at their level. Because of my own experience" (Yr. 4). Another student comments: "The first one that springs to mind which I believe that I've got and which I think is very important particularly in maths as well, would be patience" (Yr. 4). A new teacher is more expansive:

> Well I'm sensitive towards children who might have difficulty with maths because I know how it might feel and I don't want children to not feel confident with maths ... I use an encouraging and positive approach with them and ... because I think if you're struggling in maths the last thing you want is your confidence being knocked in it, you want someone to use different strategies in trying to explain something to you and use a very positive, encouraging approach and not make the child feel quite—Oh they can't do maths never ... you know, so, yeah, I think my own experience in maths has allowed me to use a certain approach with children.

Such happy resolutions to the skills required to teach mathematics can provide effective masks to the continuing anxieties relating to the students' own mathematical abilities. The evidence in our interviews pointed to such anxieties being sidestepped rather than removed since they were still apparent in relation to more explicitly mathematical aspects of our inquiry.

How then might we better understand the teachers' task of their own professional development? Professional development has it seems now come to be seen in terms of better achievement of curriculum objectives as framed within the National Numeracy Strategy (DfEE, 1999). The new teachers seemed very comfortable with this Strategy as an approach to organizing practice, even if many did find it very prescriptive. The Strategy does seem to have provided a language that can be learnt and spoken by most new teachers interviewed. In this sense, the official language spanning the National Numeracy Strategy and the inspectorial regulation of this seemed to be a huge success. This does, however, point to a need to find ways of adopting a critical attitude in relation to the parameters of this discourse in that certain difficult issues are being suppressed rather than removed. For example, when confronted with mathematics of a more sophisticated nature within the school curriculum, the new teachers remained anxious. The National Numeracy Strategy and college training however had between them provided an effective language for administering mathematics in the classroom in which confrontation with more challenging aspects of mathematics could be avoided. If the Strategy and college training had this effect, then what becomes apparent are certain limits in the teachers' capacity to engage creatively with the children's own mathematical constructions. We set these constraints up against the enabling effects of students' efforts to close gaps in their personal mathematical knowledge. Perhaps further professional development in mathematics education for such teachers might be conceptualized in terms of renegotiating these limits.

Policy initiatives are surely designed to promote improved practice. Actual improvements however may transcend the conceptualizations embedded within the initiatives. It is important to keep alive the debates that negotiate the boundaries of mathematical activity in the classroom and how those boundaries might reshape in response to broader social demands. It would be unfortunate if the prevailing conception of teacher development reached further toward the preference of providing a new set of rules, and of the teachers understanding their own professional development in terms of following those rules more effectively.

Trainees and teachers seem to be increasingly *interpellated* by multiple discourses and risk ending up speaking as if they were ventriloquists' dummies. Immersed in socially acceptable ways of describing their own practice, the obligation to identify with these ways can generate resistance to the desire to produce an identity of their own. School mathematics, as it is presently conceived, seems to have a habit of deflecting people from creative engagement into more rule governed behavior. Such moves into more structured territory could perhaps be seen as an attempt to dampen the emotional difficulties activated by attempts at more autonomous activity. It seems important, however, that further professional development be seen in terms of teachers seeking to recover and then develop some sense of their own voice toward participating more fully in their own professional rationalizations. Effective implementation of the National Numeracy Strategy is one thing. But we do need to guard against this restricting the teachers' need and desire to reconceptualize and develop their practice in an increasingly sophisticated language.

Very often, research focused on mathematics education is seen from the external perspective of mathematics experts detecting the formation of mathematics in classrooms or from the perspective of government officials concerned with administering schools and the standards they achieve. In a professional environment increasingly governed through ever more visible surveillance instruments, such as high profile school inspections, there is a sense of needing to be what one imagines the Other wants you to be, in an environment of supposed or intended control technology. But does it have to be an environment such as this? If there is a shift of focus, where more attention is given to the perspective of the emotionally charged and vocal individual teacher at the center of the classroom, development within classroom practices can perhaps be conceptualized more by those within the classrooms. Such a shift, in our view, would be positive.

ACKNOWLEDGMENT

This chapter reports on work carried out within two projects funded by the UK Economic and Social Research Council: Primary Student Teacher's Understanding of Mathematics and its Teaching (R000222409) and The Transition from Student to Teacher of Mathematics (R000223073). We would also like to thank project team members Olwen McNamara, Una Hanley, Tehmina Basit and Lorna Roberts. We thank the editor of the journal *Research in Education* for permission to reproduce some extracts from an earlier article.

REFERENCES

Bibby, T. (2002). Shame: An emotional response to doing mathematics as an adult and a teacher. *British Educational Research Journal, 28*(5), 705–721.

Brown, T. (2001). *Mathematics education and language: Interpreting hermeneutics and poststructuralism* (Revised Second Edition). Dordrecht: Kluwer.

Brown, T., & Jones, L. (2001). *Action research and postmodernism: Congruence and critique.* Buckingham: Open University Press.

Brown, T., McNamara, O., Hanley, U., & Jones, L. (1999). Primary student teachers' understanding of mathematics and its teaching. *British Educational Research Journal, 25*(3), 299–322.

Convery, A. (1999). Listening to teachers' stories: Are we sitting too comfortably? *International Journal of Qualitative Studies in Education, 12*(2), 131–146.

Davies, B. (1989). *Frogs and snails and feminist tales: Pre-school children and gender.* Sydney: Allen & Unwin.

Department for Education and Employment. (1999). *National numeracy strategy: Framework for teaching mathematics from reception to year 6.* Sudbury: DfEE Publications.

Giddens, A. (1991). *Modernity and self-identity.* Cambridge: Polity Press.

Haug, F. (1987). *Female sexualization* (Trans: Erica Carter). London: Verso.

Jones, L., Brown, T., Hanley, U., & McNamara, O. (2000). An inquiry into transitions: Moving from being a learner of mathematics to becoming a teacher of mathematics. *Research in Education, 63,* 1–10.

Kagan, D. (1990). Ways of evaluating teacher cognition: Inferences concerning the Goldilocks principle. *Review of Educational Research, 60*(3) 419–469.

Laclau, W., & Mouffe, C. (2001). *Hegemony and socialist strategy.* London: Verso.

Lather, P. (1991). *Getting smart: Feminist research and pedagogy with/in the postmodern.* London: Routledge.

MacLure, M. (1993a). Arguing for yourself: Identity as an organizing principle in teachers' jobs and lives. *British Educational Research Journal, 19,* 311–22.

MacLure, M. (1993b). Mundane autobiography: Some thoughts on self-talk in research contexts. *British Journal of Sociology of Education, 14*, 373–384.

Munby, H. (1986). Metaphor in the thinking of teachers: An exploratory study. *Journal of Curriculum Studies, 18*, 197–209.

O'Connell Rust, F. (1999). Professional conversations: New teachers explore teaching through conversation, story and narrative. *Teaching and Teacher Education, 15*, 367–380.

Ricoeur, P. (1981). *Hermeneutics and the human sciences.* Cambridge: Cambridge University Press.

Scheff, T.I. (1994). *Bloody revenge: Emotions, nationalism and war.* Boulder, CO: Westview Press.

Schon, D. (1979). Generative metaphor: A perspective on problem setting in social policy. In A. Ortony (Ed.), *Metaphor and thought.* Cambridge: Cambridge University Press.

Stronach, I., & MacLure, M. (1997). *Educational research undone: The post-modern embrace.* Buckingham: Open University Press.

Stubbs, M. (1983). *Discourse analysis: The socio-linguistic analysis of natural language.* Blackwell: Oxford.

Walkerdine, V. (1988). *The mastery of reason.* London: Routledge.

Walkerdine, V. (1990). *School girl fictions.* London: Verso.

Walshaw, M. (2001). A Foucauldian gaze on gender research: What do you do when confronted with the tunnel at the end of the light? *Journal for Research in Mathematics Education, 32*(5), 471–492.

Zizek, S. (1989). *The sublime object of ideology.* London: Verso.

CHAPTER 10

SO WHAT'S POWER GOT TO DO WITH IT?

Tamsin Meaney

ABSTRACT

As a non-indigneous person working with a Māori school community to develop a mathematics curriculum, issues of power and control were never far from my mind. Using ideas from Michel Foucault, in this chapter I reflect upon some of the interactions that occurred between myself and members of the school community. In particular I document how the process of determining what knowledge was accepted as valuable was shared between us even in situations where it appeared that I controlled the interactions. I conclude with possible implications for non-indigenous educators working with indigenous communities.

INTRODUCTION

Thinking critically is not necessarily a natural occurrence. It doesn't automatically arise simply because one is told to look for problems. Rather, such an awareness is built through concentrated efforts at a relational understanding of how gender, class, race, and power actually work in our daily practices and in the institutionalized structures we now inhabit. (Apple, 1992, p. 418)

Mathematics Education Within the Postmodern, pages 181–199
Copyright © 2004 by Information Age Publishing
All rights of reproduction in any form reserved.

Apple reminds us that thinking critically about issues such as disadvantage within schooling systems requires more than an identification of problems. What is needed instead is an investigation into the complexity of such issues and how beliefs held by certain sections of society have determined which behaviors and practices are acceptable ones while the rest remain those of disadvantaged peoples or Others (Klages, 1997). Postmodernism offers ways to critique these beliefs and how they are formed. A micro-level analysis of actual events and self-reflection by those involved provides insight into the complexity that makes up these events.

This chapter describes a postmodern analysis. I draw on the work of Foucault to analyze the interactions between myself (a non-indigenous educator) and an indigenous community, during a community-negotiated mathematics curriculum development project. I explore how the power, which ebbed and flowed within the professional relationship between myself and community members, was affected by two other types of relationships (social and societal). In scenarios based on examples of different behaviors, I describe fluctuations of power through the ways in which the community engaged with and utilized what I offered in my capacity as mathematics educational consultant.

Michel Foucault's ideas were crucial in this investigation as he believed that power "needs to be considered as a productive network which runs through the whole social body, much more than a negative instance whose function is repression" (Foucault, quoted in Gordon, 1980, p. 119). In trying to understand what was happening in my own interactions there was a need to move, as Apple suggested, beyond identifying problems and apportioning blame to discovering more. Foucault's ideas about power enabled me to think critically by gaining an awareness of how both the community members and I were positioned in relation to each other as the project progressed. This has since provided me with greater awareness of how I might operate in similar situations. Through my description of the methods used in this self-reflection, it is hoped that they will be available to others for adaptation to their own areas of inquiry.

THE RESEARCH EVENT

The research examined how parents and teachers of a New Zealand Māori immersion school (Kura Kaupapa Māori) used a resource document (*The Framework*) produced by me to support the development of a mathematics curriculum. The pedagogy within Kura Kaupapa Māori is based on Māori preferred teaching and learning methods and parents at these schools are expected to take an active role in school decision making. The Framework provided background and discussion questions on several aspects of math-

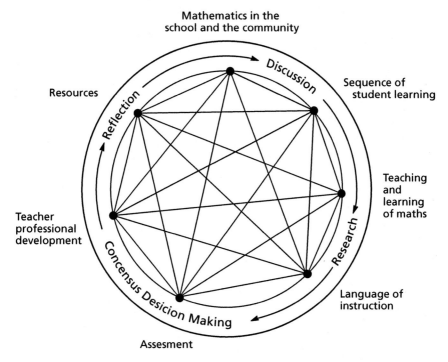

Figure 10.1.   An outline of the issues and method suggested in the Framework

ematics curriculum development and proposed a method for investigating these issues. In Figure 10.1 (from Meaney, 1999, p. 86), the aspects are distributed around the outside and the method is described inside the outer circle. The lines joining the nodes illustrate that curriculum development is not a linear process but rather one where decisions made on one issue will affect all other issues.

As the project progressed, what came to the fore was the issue of the role of an outsider within indigenous education. In cross-cultural situations, the power which can be channeled through the knowledge of the educational consultant is caught up in the culture and history of the groups consulted. At the beginning of the project, it was anticipated that insights about the consultant's role could be gained by analyzing feedback about the Framework. However, although much data were gathered from nine community meetings and from interviews throughout the project, with the parents and teachers involved, there were very few comments about the Framework or about my role (see Meaney, 2001a). It was obvious that a different approach had to be found if the issue was to be addressed adequately.

POWER AND RELATIONS

There has been much discussion about the roles that people can play in projects oriented toward social justice. For example, Freire (1996) divided people into oppressors and the oppressed and was emphatic that oppressors were unable to participate in the reflection and action process that constitutes praxis. Darder (1991) distinguished between those from dominant and subordinate cultures. These distinctions, although helpful in highlighting differences, did not seem adequate in facilitating understanding about the impact of the role that I performed. Knijnik (2000) in discussing the work of Popkewitz argued that dividing the world into oppressors and oppressed "cover[s] up the actions and practices of individuals through which power also operates" (p. 5).

Foucault's ideas seem relevant for an understanding of my role and for assessing whether or not my actions replicated the traditional role of dominance of people from a mainstream background over those from an indigenous background. For Foucault, power could be ingrained in the interactions of an outside consultant, because using and producing knowledge in conjunction with a community is an integral part of the consultant's role. Foucault believed that "the exercise of power itself creates and causes to emerge new objects of knowledge and accumulates new bodies of information ... and, conversely, knowledge constantly induces effects of power ... It is not possible for power to be exercised without knowledge, it is impossible for knowledge not to engender power" (Gordon, 1980, pp. 51–2).

In this way, power can produce new, or validate existing, knowledge that comes to be seen as the truth or the reality about a situation (Gordon, 1980). We can see this operating within mathematics education: Zevenbergen (1996) claims that because the abstract qualities of mathematics have been valued so highly by society, the passing on of this knowledge and way of thinking has been rigidly controlled. Therefore, mathematics education has concentrated on improving students' chances of learning mathematics rather than improving students' ability to critique the role mathematics plays in society.

Underpinning the curriculum development project was an assumption that discussions about mathematics and school achievement could help to make explicit the mismatches between the education system and the indigenous community about what knowledge should be considered valuable or the "truth." This had been highlighted as important by others such as Lipka (1994) and Cantoni (1991) who have worked in curriculum development with indigenous communities.

Cultural negotiation is a process that makes schools' hidden values and processes visible to community and school while making the community's knowledge, values, and processes visible to schooling. Schooling then becomes explicit and open to choices—choices that can only be responded to at the local community level as they concern issues of culture, language, and identity. Through an exploration of their own cultural strengths and their particular goals and visions for their children, community and school can construct a curriculum of the possible—creatively devising content and pedagogy. (Lipka, 1994, p. 27)

The role of an outside expert within such curriculum projects is unclear. Many overseas aid projects are not funded unless an outside consultant is involved. However, without an analysis of power relations, experts' roles can be rejected as having nothing to offer because they are seen as "instruments of an all-powerful establishment" (Schon, 1983, p. 10). In cross-cultural work and research, the issue of power is often raised (e.g., Bishop, 1996; Smith, 1998). There is anecdotal evidence to suggest that the fear of imposing knowledge upon another group can result in no knowledge being provided. For example, Delpit (1993) in her discussion of the ideas of James Paul Gee regarding the teaching of dominant discourses to people of color, wrote that some teachers become reluctant to do so, in case replacing students' home voices resulted in those voices being further oppressed. On the other hand, she suggested that to avoid teaching the dominant discourse is also oppressive, as these students "need access to dominant Discourses to (legally) have access to economic power" (p. 292). By being fluent in the dominant discourses, they have the ability to transform those discourses. This suggested that if I did not make my knowledge available to indigenous communities, those communities would be disempowered.

The ultimate aim of my work was to improve indigenous students' achievement in mathematics education. This, however, needed to be carried out *with* the community not *for* them (see Osborne, 1995). If *I* decided on the criteria by which improvement could be achieved and the problems that needed to be overcome for this to happen, then regardless of my good intentions I would not be benefitting indigenous students. Instead the possibilities available to indigenous communities about the mathematics learning of their children would be limited. I would be participating in hegemonic evangelism as described by Bishop (1996). In this situation "[t]he research relationship would remain paternalistic and hegemonic despite the degree of the concern of the facilitator" (p. 59). In order to overcome this, I needed to "ensure that the analysis that the intellectual her/himself performs of her/his role in the social world contain[s] a larger dose of self-reflexiveness on relations between knowledge, power, intellectuals, teaching and research" (Knijnik, 2000, pp. 4–5).

ANALYZING POWER

In reviewing the data from the meetings and interviews, I had to decide how to use Foucault's ideas to understand what had happened. Researchers, such as Gore (1998) use Foucault's work to identify techniques of power that are activities leading to the construction of reality or truth (Hardy & Cotton, 2000). Gore, in her investigation of four very different educational situations, searched for the eight techniques of power that Foucault had identified as operating within penal institutions. She was able to find all the techniques at each site (although some were used more extensively than others). However, by looking for particular techniques of power, it seemed to me that the cohesiveness of the situation became fragmented and the ways that power both produces and represses knowledge remained opaque.

I chose to investigate the determining effects of power and to explore some of the ways that knowledge and truth are produced within a series of scenarios. I classified and examined three levels of relationships. Within those levels, the situations remain intact, and the (sometimes conflicting) effects of the techniques of power could be clarified.

THREE LEVELS OF RELATIONSHIPS

Foucault (in Gordon, 1980) stated that "[i]n reality power means relations, a more-or-less organized, hierarchical, co-ordinated cluster of relations" (p. 198). It therefore seemed important to consider how different relationships impacted on each scenario. The primary relationship that I had with the community in the project was professional. However, both social and societal relationships influence the operations of a professional relationship. These are therefore defined before describing the model of behaviors and the scenarios from the project.

Professional relationships and my actions within them were connected to the expectations and needs of the community with whom I worked. In this project, I provided support, mostly through the Framework, to a school community who were considering developing a mathematics curriculum. In the first place approval from the school board for my involvement was necessary and was obtained in due course through the auspices of the principal.

In a professional relationship, it is possible that within the community my knowledge would take precedence because of its status as belonging to the consultant. The flow of power within this relationship could be considered less fluid as a consequence. Within a social relationship, it is more likely that if one person's needs and interests seem to monopolize, the others involved will either confront that person or withdraw from the relationship. In professional relationships, it may not be possible to withdraw and the aura of expertise may inhibit other participants in the project from confronting the consultant.

Social relationships are those such as friendships that develop from knowledge of people. Within a friendship, whose knowledge controls the situation will depend upon whose interests are deemed as more important. However, it is expected that the trust within a friendship ensures that no one person's interests will dominate all the time.

Some of the people with whom I engaged in the project had been friends before the meetings began. Some regular attendees also began to become friends. This meant that we often had conversations about children and other interests which we had in common. Relationships of trust with community members were essential for the research component of this project. However, my attempts at building this trust were not always successful. Living in another city from the community meant that, during the course of the project, I attended only one function that was not directly related to the curriculum development meetings or interviews. Without opportunities to form social relationships, I was unable to develop the necessary mutual trust and this had considerable impact on the research.

Societal relationships describe the positioning of different groups within society as a consequence of the development of cultural norms over long periods of time. For instance, Cummins (1996) related how First Nations students, from Canadian indigenous communities, were treated in residential schools:

> "[T]he interactions between individual educators and students (henceforth termed micro-interactions) were merely reflecting the pattern of interactions between dominant and subordinate groups in the wider society (henceforth macro-interactions) where First Nations communities were widely disparaged. In both micro- and macro-interactions, the process of identity negotiation reflects the relations of power in the society. (p. 15)

Societal relationships allow people to operate without having to consider the implications of every action or thought because their beliefs are ingrained as true representations of how the world is. This, however, makes societal relationships difficult to recognize and harder to change. McIntosh (1988) explored the parallels between white privilege and male privilege, both of which could be considered as being influenced by societal relationships. She wrote:

> I think whites are carefully taught not to recognise white privilege as males are taught not to recognize male privilege … I have come to see white privilege as an invisible package of unearned assets which I can count on cashing in each day, but about which I was "meant" to remain oblivious. (pp. 1–2)

The dominant group sees its knowledge as the most valuable not only because of its own beliefs but also because those who are in the subordinate positions have also accepted this reality. The value of this knowledge remains unquestioned because to suggest that any other possible situation is similar to questioning your own existence. For example, McIntosh (1988) discussed how "whites are taught to think of their lives as morally neutral, normative and average, and also ideal, so that when we work to benefit others, this is seen as work which will allow 'them' to be more like 'us'" (p. 4). This is how the world is, rather than something that could be challenged and changed.

My position as a well-educated, middle-class, non-indigenous professional influences all of my interactions. Although as a consultant, part of my role requires me to facilitate others to query the accepted truths about their situations, my own cultural background may not make it possible to see these truths as questionable. A later section in this chapter examines, in detail, how my membership of this group influenced how I performed my professional role.

SCENARIOS

Figure 10.2 depicts eight examples of behavior that I could engage in as a consultant working in a community. I felt that it was important to recognize that even the best intentioned consultant is likely to use, what might at first appear to be, inappropriate behaviors. It was therefore important to acknowledge this first and then to investigate how power circulates when such behaviors are used. The model is based very loosely on Wubbels' model of teacher interpersonal behavior (cited in Fisher & Rickards, 1998, p. 5). The scenarios illustrate situations where power often fluctuated between myself and other participants as different knowledge was used and developed. Both the influences affecting the availability of knowledge and the interactions within and among the three types of relationships are examined.

The model has two axes: Control and Attitudes to Action. The behaviors exhibited on the Control continuum, represented on the vertical axis, vary between dominance and collaboration. Dominance is where, through either "paternalistic" or "oppressive" behaviors, I might limit the choices that the school community could make regarding actions or knowledge. Collaboration, on the other hand, has been described as containing "mutual respect for skills and knowledge, honest and clear communication, two-way sharing of information, mutually agreed upon goals, shared planning and decision making" (Christenson, Rounds, & Franklin, 1992, p. 20). Located at this end of the vertical axis, my behavior could be "participatory" and supportive or "exploratory."

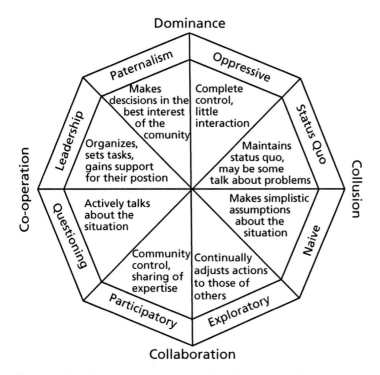

Figure 10.2.   Model of behaviors exhibited by an outside consultant.

The behaviors exhibited on the attitudes-to-action continuum, which lie on the horizontal axis, move between collusion and co-operation. Collusion at the far right of the horizontal axis, was described by McDermott (cited in Osborne, 1989) as "maintaining the status quo in the social and political relationships between two cultures" (p. 197). Collusion allows for knowledge to be questioned but in such a way that underlying assumptions are not challenged and so the situation remains unchanged. Within collusion behavior can either maintain the "status quo" or be "naïve." Co-operative behaviors, at the extreme left, are "joint efforts of a number of people that are mutually beneficial to all involved parties" (Dunst, Johanson, Rounds, Trivette, & Hamby, 1992, p. 157). They are represented by "leadership" and "questioning" behavior and contribute to changing situations.

To clarify how the eight behaviors operated in my curriculum development project I describe each in turn, illustrating with examples from the data. An analysis is then made of how knowledge was produced and how power was used.

Paternalistic behavior is exhibited when decisions are made by the consultant without discussion but "in the best interests" of the community

(Ellis, 1976). I wrote the Framework for a hypothetical community who would need support in order to participate in mathematics curriculum development. The actual community was not consulted in any depth until two months before the first meeting and in the interviews from this time the idea of using a support document such as the Framework was not discussed. Even though the Framework did not fulfill the function that I anticipated, I had difficulty giving up the idea that it was an appropriate form of support for communities who want to develop a mathematics curriculum.

As the consultant, I believed that I had developed, for indigenous communities, a creative solution to a problem that I identified—that of communities not having sufficient curriculum development expertise. I came to see the Framework as a way that any community, anywhere, could be provided with background information and discussion questions to facilitate their curriculum development. I accepted the need for such a document. The decision to provide support in such a format was never queried by the school community and by accepting this decision they accepted my control. In setting the project up with the Framework as nonnegotiable, I limited the range of options available to the community. It may have been difficult to challenge me about using the Framework because the social relationships between community members and myself were such that this could have caused conflict. If the social relationships had been stronger, the professional relationships may have been able to withstand such a challenge.

Oppressive behavior is exploitative, controlling and dominating (Ellis, 1976) and occurs when the consultant makes unilateral decisions. Although my previous work in Aboriginal communities influenced my decisions about what to include within the Framework, no indigenous people were consulted about what should be in the initial draft. My decisions constrained what the school community did as the Framework's issues and the discussion questions within them became the background for the meetings.

However, the community responded to the issues in the Framework in their own way. The discussions that I had anticipated would happen did not occur but instead other ideas were raised. At the end of the project in reflecting on what had occurred, the principal of the school described the Framework as a "soft" document.

> Because you are trying very hard not to be directive, but maybe some of those questions have to be asked. Maybe people have to be asked on a cultural continuum, how do they feel about the mainstream thing and think it out hard ... The document is really gentle. You don't ask any questions that start people bleeding straight away. (Meeting 9)

Yet, if the questions in the Framework had been made more directive, then the community might have had less control over how to interpret it. This may have resulted in the community deciding not to use the Framework because they felt uncomfortable with its approach.

Power fluctuated between us as different knowledge came to be seen as valuable. My choice of what issues to include was not questioned directly. However, the community used the discussion questions as starting points for raising their own dilemmas (see Meaney, 2001b) and so developed their own truth about the mathematics education of their children. Although my behavior appeared to make them uncomfortable, in fact my choices about what knowledge to raise did not control the outcomes from our interaction.

Supportive behavior involves sharing knowledge and expertise. Who makes decisions is not differentiated from who is affected by those decisions. In the penultimate meeting, the community considered how the issues raised in the Framework matched what was in *Te Aho Matua* (Kura Kaupapa Working Party, 1989), the philosophy document for kura kaupapa Māori. From their discussion, I identified ideas that could be included in the Framework such as the celebration of learning. At the same time, the community discovered ideas that had not been included in *Te Aho Matua* but which the community felt were important in the school's philosophy, such as professional development for teachers.

At this meeting, knowledge was shared. In this discussion, both the Framework and *Te Aho Matua* assumed equal status as the comparison was not to discover which document was better but to find out what each had to offer mathematics curriculum development. As the knowledge gained from the comparison was of use to the community and to me, so power was equally deployed in this situation. I decided what changes were made to the Framework while they controlled the outcomes of the comparison in regard to their mathematics curriculum ideas.

Exploratory behavior. Exploratory behavior occurs when different ways of interacting are tried out within situations. The issue concerning inclusion of questions about ethos was raised by the principal in the eighth meeting. He felt that indigenous communities who were in revivalist mode may have greater difficulty conceptualizing different kinds of mathematics. Although I was happy that he had raised this with the community, it did not seem appropriate to me for it to be incorporated into the Framework. In that discussion, I stated what I was and was not prepared to do. In other discussions, we explored together possibilities for providing support to the curriculum development process, including talking with the Ministry of Education and developing graphs from the results of questionnaires (see Meaney, forthcoming). On some issues, such as adding questions about the features of language impacting on the learning of mathematics, I acquiesced to their suggestions. As a result of this exploratory process the community and I made some adjustments to our expectations of my role. As the negotiation continued, power fluctuated between us. The actions of the community were not obviously constrained and different possibilities were seen as available.

Status quo.   Status quo behavior ensures that wider societal conditions are not challenged and so problems within an immediate situation, although identified, are not resolved. After one community meeting, I was asked to summarize from the audio recording the main ideas that had been raised. Large parts of this report were later included in the school's policy document on mathematics education.

Documenting their views was a very powerful action. It was felt that as an attendee at the meeting, I would have reported what was said accurately. Their trust in the documentation overcame any concerns that they may have had about my interpretation. As a non-community member, I may not have been able to recognize issues of importance to the community. My identification of other issues may have rewritten community members' memories of what had been discussed so that other issues were forgotten. The community, by accepting the report, came also to accept my use of power. The status quo was unchanged because although problems with my writing of the report were identified and relayed back to the community, the contents of the report were accepted. Thus other concerns that affected this community were not highlighted as important and so were not considered as changeable.

Naive behavior is behavior where simplistic assumptions are made and, as a result, possible actions restricted. I was naive in believing that many parents would want to be involved in the curriculum development meetings and that their views would have the same status as those of teachers. Anderson (1998) identified both these aspects as problems in participatory educational reforms. He believed that parents who were the least comfortable in a school were the least likely to attend meetings. Parents whose own education had positioned them as non-achievers were unlikely to participate in meetings to explore ways to help their children improve their achievement. In addition, he reported two studies where "participatory groups meant to foster dialogue end up producing a professional monologue that results in a 'parody of collaboration'" (pp. 580–581).

One of the principles of the Framework was to involve community members in curriculum development. By not considering carefully how to achieve this, I set up a situation in which this outcome was unlikely to eventuate. After I became aware of what was happening, I continued to be paralyzed in my interactions with community members as I made no use of this information. I felt unable to interfere with how they ran their meetings in case this caused conflict and restricted my access to interviews. In this situation, the responsibility that goes with professional relationships was not fulfilled. If I had examined my own knowledge more carefully and had offered what I knew to the community, then the power that went with this knowledge could have been shared with community members. Once again, if our social relationships had been stronger then I would have been less concerned about causing conflict.

Leadership behavior occurs when community members' support for the facilitator's organizing, task-setting, and proposals is gained. In this project co-operation was needed to carry through a course of action but this action had already been decided upon by the facilitator and so there was limited sharing of ideas.

Originally I had felt that the problems encountered in organizing meetings could be left to the community to resolve. However, when it seemed people were indicating their dissatisfaction with the all-talk-and-nothing-achieved meetings, I decided to intervene. Although I talked with parents and teachers in interviews about the situation, it was simply to see whether they thought that a particular response was feasible. My knowledge of the community and how it operated was limited and so a response was suggested based on what I thought had previously operated in this community. This solution was not effective in resolving the problem. Community members' uncertainty about the curriculum development process and mathematics meant that most were not confident to take on the roles that I proposed for them.

In this situation, I came to see my knowledge as more useful than it was. I used that knowledge and the power I thought it contained, to gain support for my ideas from the community. Although the community agreed to my organizational ideas, their subsequent actions suggested that they had not considered them appropriate. They used their knowledge of themselves to passively resist my suggestions rather than to confront me directly. It may well have been that there were other layers to this problem that my limited knowledge of the community did not let me recognize. Spending more time with the community may have provided me with a better understanding of the situation.

Questioning behavior is when a facilitator questions community members about their views but makes only limited use of that information. After the first meeting I was concerned that if the community continued to label everything as mathematics, then the usefulness of such a label would be lost. Without the community discussing the implications of this decision, I felt that they would be unable to make an informed decision in regard to the mathematics curriculum. Therefore, I made several attempts to raise this issue in different ways, including directly questioning community members within meetings and in interviews. Discussion about this issue had still not occurred by the end of my involvement with the project. I interpret their actions as exerting their right not to discuss an issue that to them seemed irrelevant. Instead I note that they made use of the opportunities that I provided for these discussions to discuss other things. As the consultant, I would have been negligent in not raising the issue but I needed to accept that the community did not share my concern. Thus power circulated between us as knowledge was offered, rejected and used to develop other knowledge.

THE USE OF POWER/KNOWLEDGE BY AN OUTSIDE CONSULTANT

The scenarios illustrate how knowledge was used both by the community and by me to produce mathematics curriculum for particular situations. In this way, power circulated in the relationships between us. Although sometimes the school community accepted my definitions of curriculum, there were other situations in which the community rejected both my posing of problems and the solutions that I offered. In these situations, community members either explained to me what their reasons were for their rejection or they accepted my proposals at first but later chose not to act upon them. Similar situations arose when community members defined problems and solutions and I accepted or rejected them. There were also times when the very meaning of curriculum was negotiated. When my behavior was supportive or exploratory, we came to conclusions together that we found mutually acceptable. Foucault's belief in power as dispersed, indeterminate, productive, dynamic, was evidenced in this analysis. These examples suggest that although at particular points in time, my apparent expert knowledge controlled the outcomes of a situation, this was not always the case. It was only when I set up a situation as accepted practice that the options for community control became extremely limited. This is what happened when I discussed the project with the school community with the Framework as a nonnegotiable part of my involvement.

What does this have to do with how I performed my role? Freire (1996) contends that for learning to occur it must happen within dialogue based on love, humility, faith, trust, hope and critical thinking. If these attributes of dialogue are extended to people involved in mathematics curriculum development, then an outside consultant should always exhibit collaborative behavior. If the consultant uses participatory and exploratory behaviors, then negotiation becomes essential and provides the community with more opportunity to control what is being discussed and the outcomes of those discussions. In this way the consultant is a facilitator rather than an overseer.

However, in situations where my work is with a community whom I do not know very well, I learnt that my behavior is more likely to be dominating. By knowing more about a community (what they know, what they want and what they are prepared to do), negotiation about the consultant's role becomes easier. In project work on communities, however, time restrictions often mean that it is not possible to discuss the implications of all the options available. The less that is known about the community, the more the consultant relies upon their previous experiences to gauge what should be done. Although the community would still have ways to exercise power, including withdrawal, in their view, the value of the consultant's role is minimized.

Most projects undertaken by outsider consultants are carried out with little initial knowledge about the community. Dominating behaviors come to be seen as the only option because collaborative behaviors require negotiation through sharing of expertise and aims. This takes time and a degree of trust on all sides. The outcomes of professional relationships are influenced by both social relationships and societal relationships. In several of the scenarios, I felt that better social relations would have enabled more sharing about the situation to take place. As a consequence, the usefulness of projects which are carried out without initial knowledge of the community must be queried—not so much for what they do achieve but for what they may have achieved if those social relationships of trust had been developed originally.

Even when there is negotiation about a project, there may be times when dominating behaviors are appropriate for the consultant. Sometimes, the consultant's knowledge will determine a particular course of action. However, there need to be restrictions about when dominating behaviors can be used. A consultant may make decisions about how to facilitate a process but the outcomes of the process itself needs to be left to the community to decide. Although there will still be problems with a consultant making decisions of this kind, as was illustrated in the scenarios, there will also be opportunities for the school community to modify the situation.

THE INFLUENCE OF SOCIETAL RELATIONSHIPS

The work of Apple (1992) and others raised my levels of awareness regarding the acceptance of the dominant group's beliefs and how those beliefs have contributed to the "failure" of others. Societal relationships made it difficult to recognize that my beliefs were merely assumptions. In fulfilling my role, I needed to ask in what ways have my actions held intact the wider power relations within society? From this research, there were some clear examples of where my acceptance of societal taken-for-granted under-standings had limited my role as consultant. These included allowing the Western schooling system to remain unquestioned within the project. Although I am aware that there are some gaps within the present educa-tion system, I decided not to raise this as an issue within the Framework. Societal relationships, therefore, affect the suggestions made by a consult-ant by limiting what the consultant considers as possibilities. This in turn can restrict the types of decisions that can be opened up for communities to consider.

At times, this research has caused me great anguish as I have begun to pull apart some of this accepted knowledge about societal relationships. The level of self-reflexiveness required in postmodern research means that to be useful some of the things that we feel make up who we are, such as our role in society, need to be critiqued and will be found wanting. How-ever, in order for the situation within education to be changed these understandings must be challenged. My professional relationships will be of limited effectiveness if I am unable to recognize the impact of societal relationships upon what I do.

The question, then, becomes one of how to challenge accepted knowl-edge and the power that it contains. When I began the research, I wrote down my assumptions about the Framework, who would use it and how it would be used. Later, this clarified my own beliefs concerning mathematics curriculum development in communities. I came to see more clearly my beliefs about indigenous communities, teachers and the mathematics cur-riculum and about my own role. These gave me insights into my beliefs about schooling, education and society—not a theoretical understanding from the literature but an understanding of what I had come to accept as the truth about these things. If I were to use the Framework with another community, my assumptions could be used as starting points from which to negotiate my role within a project.

CONCLUSION

This chapter examined some scenarios from a curriculum development project and looked at how power operated in the acceptance and valida-tion of knowledge. Foucault in his analysis of the development of bodies of knowledge within the human sciences wrote extensively about the connec-tion between power and knowledge (Gordon, 1980). One of his central ideas was that power fluctuates between people as knowledge comes to be accepted. In some of the scenarios, I have portrayed power as ebbing and flowing between community members and myself as different knowledge was offered, produced, modified and accepted. However, it is suggested that in projects where there is little initial knowledge about a community, a consultant is more likely to use dominating behaviors thus limiting a community's chances of controlling the exercise of power through the negotiation of meaning. In order for collaborative behaviors to occur, there needs to be a sharing of knowledge and purposes so that negotiation of the consultant's role can occur. These behaviors can only develop from the trust built up through social relationships over time. In order for a project to be effective, an outsider needs to ensure that adequate time is set aside to build up social relationships.

Societal relationships also influenced many of the interactions that I had with community members. As what becomes accepted as "true" becomes ingrained through these societal relationships, the influence of the relationships is much harder to identify. Alternative actions are not apparent because the need for them is not recognized. This often unacknowledged power, embedded within these relationships, has considerable control over what occurs. Without a questioning of the knowledge accepted through these relationships, I became an unwitting vehicle for power regardless of my good intentions. It was only after much reflection during the course of this project that I became aware of my own embedded assumptions. It now seems that in this project I did determine a problem and also the solution, and hence participated in hegemonic evangelism. The project's saving grace was that there was enough flexibility incorporated within it for the school community to exert their control and so develop a discussion of the issues they considered important. The community also gave me leeway so that although what I offered them did not always match their needs, they did not reject it but rather used it in a way that better suited their purposes. Our working together could be read as an offering of information that was rejected, accepted or modified as power circulated between us.

So, what has power got to do with it? Power is inherent within all interactions. As a professional, potentially my role could control situations. Although this did not always appear to be the case in the scenarios examined, the acceptance of some knowledge as truth at the societal level suggests that both my actions and the actions of the community members were constrained. However, as the consultant with these professional relationships, I was responsible for raising issues with the community so that they were able to make their own decisions about the mathematics education for their children.

Foucault's beliefs about the circulatory nature of power within a hierarchical set of relations provided insight into the interactions between the community members and me. In this research a model was used which labeled particular incidents according to their surface features. However, it was only by teasing these incidents out, by moving away from superficial labels, and by engaging in considerable reflection, that insight was gained. By looking at how people are positioned through the production, acceptance and validation of particular knowledge, I gained a greater understanding of the complexity of the issues involved in cross-cultural curriculum development. Similar understandings can be sought by others in other situations where differentials in status exist between participants, such as within classrooms. Using a model, such as the one suggested here, may allow others to begin to comprehend and analyze the complexity of the educational process.

REFERENCES

Anderson, G.L. (1998). Toward authentic participation: Deconstructing the discourse of participatory reforms in education. *American Education Research Journal, 35*(4), 571–603.

Apple, M.W. (1992). Do the standards go far enough? Power, policy, and practice in mathematics education. *Journal for Research in Mathematics Education, 23*(5), 412–431.

Bishop, R. (1996). *Collaborative research stories: Whakawhanaungatanga.* Palmerston North: Dunmore Press.

Cantoni, G. (1991). Applying a cultural compatibility model to the teaching of mathematics to indigenous populations. *Journal of Navojo Education, IX*(1), 33–42.

Christenson, S.L., Rounds, T., & Franklin, M.J. (1992). Home-school collaboration effects, issues and opportunities. In S.L. Christenson & J.C. Conoley (Eds.), *Home school collaboration: Enhancing children's academic and social competence* (pp. 19–51). Bethesda: MD: National Association of School Psychologists.

Cummins, J. (1996). *Negotiating identities: Education for empowerment in a diverse society.* Ontario, CA: California Association for Bilingual Education.

Darder, A. (1991). *Culture and power in the classroom: A critical foundation for bicultural education.* London: Bergin & Garvey.

Delpit, L.D. (1993). The politics of teaching literate discourse. In T. Perry & J.W. Fraser (Eds.), *Freedom's plow* (pp. 285–295). New York: Routledge.

Dunst, C.J., Johanson, C., Rounds, T., Trivette, C.M., & Hamby, D. (1992). Characteristics of parent-professional partnerships. In S.L. Christenson & J.C. Conoley, (Eds.), *Home school collaboration: Enhancing children's academic and social competence* (pp. 157–174). Bethesda: MD: National Association of School Psychologists.

Ellis, H.G. (1976). Theories of academic and social failure of oppressed black students: Source, motives and influences. In C.J. Calhoun & F.A.J. Ianni (Eds.), *The anthropological study of education* (pp. 105–126). The Hague: Mouton Publishers.

Fisher, D., & Rickards, T. (1998). Associations between teacher-student interpersonal behavior and student attitude to mathematics. *Mathematics Education Research Journal, 10*(1), 3–15.

Freire, P. (1996). *Pedagogy of the oppressed* (Trans: M.B. Ramos). London: Penguin Books.

Gordon, C. (Ed.) (1980). *Power/Knowledge: Selected interviews and other writings, 1972–1977, Michel Foucault* (Trans: C. Gordon, L. Marshall, J. Mepham, & K. Soper). New York: Harvester Wheatsheaf.

Gore, J.M. (1998). Disciplining bodies: On the continuity of power relations in pedagogy. In T. Popkewitz & M. Brennan (Eds.), *Foucault's challenge: Discourse, knowledge and power in education* (pp. 231–251). New York: Teachers College Press.

Hardy, T., & Cotton, T. (2000). Problematising culture and discourse for maths education research: Tools for research. In J.F. Matos & M. Santos (Eds.), *Proceedings of the Second International Mathematics Education and Society Conference* (pp. 275–289). Lisbon: Centro de Investigação em Educação da Faculdade de Ciências, Universidade de Lisboa.

Klages, M. (1997). *Postmodernism.* Retrieved February 18, 2003, from: http://www.colorado.edu/English/ENGL2012Klages/pomo.html

Knijnik, G. (2000). Re-searching mathematics education from a critical perspective. *Mathematics Education and Society Conference 2.* Retrieved April 16, 2003, from http://correio.cc.fc.ul.pt/~jflm/mes2/programme.html.

Kura Kaupapa Maori Working Group (1989). *Te Aho Matua* Wellington: Ministry of Education.

Lipka, J. (1994). Culturally negotiated schooling: Toward a Yup'ik mathematics. *Journal of American Indian Education, 33,* 14–30.

McIntosh, P. (1988). *White privilege and male privilege: A personal account of coming to see correspondences through work in women's studies.* Working Paper 189, Centre for Research on Women, Wellesley College, Wellesley, MA.

Meaney, T. (1999). Mathematics curriculum development in indigenous communities. In A. Rogerson (Ed.), *Proceeding of the International Conference on mathematics education into the 21st century: Societal challenges, issues and approaches* (Vol. 2, pp. 85–94). Cairo.

Meaney, T. (2001a). *An ethnographic case study of a community-negotiated mathematics curriculum development project.* Unpublished Ph.D. thesis. Auckland: University of Auckland.

Meaney, T. (2001b). Parents and teachers doing mathematics curriculum development. *Mathematics Education Research Journal,* 3–14.

Meaney, T. (forthcoming). The fly on the edge of the porridge bowl: Outsider research in mathematics education. In P. Valero & R. Zevenbergen (Eds.), *Researching the socio-political dimensions of mathematics education: Issues of power in theory and methodology.* Dordrecht: Kluwer.

Osborne, A.B. (1989). Insiders and outsiders: Cultural membership and the micropolitics of education among the Zuni. *Anthropology and Education Quarterly, 20,* 196–215.

Osborne, B. (1995). *Indigenous education: Is there a place for non-indigenous researchers?* Paper presented at the Australian Association for Research in Education Conference. Hobart, Australia.

Schon, D. (1983). *The reflective practitioner.* New York Basic Books.

Smith, L.T. (1998). *Decolonizing methodologies: Research and indigenous peoples.* Dunedin: University of Otago Press.

Zevenbergen, R. (1996). Constructivism as a liberal bourgeois discourse. *Educational Studies in Mathematics, 31,* 95–113.

CHAPTER 11

WHY MATHEMATICS?

Insights from Poststructural Topologies

M. Jayne Fleener

ABSTRACT

Do we ever think about what we don't think about? For that matter, do we ever think about what we DO think about? What are the limits to our thinking? How can we expand those limits and shift those borders? In this chapter I provide a context for exploring and dissolving boundaries and for asking fundamental "why" questions. Drawing on the decentering process of Deleuzian poststructuralism I note the conundrums of mathematics and mathematics education in current curricular contexts, and offer insights about the possibility of rethinking our curriculum futures.

INTRODUCTION

In the movie *Inherit the Wind*,[1] the exasperated Spencer Tracy character, depicting Clarence Darrow in the Scopes Monkey Trial, queries the William Jennings Bryan character in a powerful scene that exposes the vulnerabilities and limits of religious dogma. He asks: "Do you ever think about

Mathematics Education Within the Postmodern, pages 201–218
Copyright © 2004 by Information Age Publishing
All rights of reproduction in any form reserved.

what you *don't* think about?" By challenging unquestioning and by exposing unquestioned assumptions, the Spencer Tracy character challenges all of us to reflect on the limits of our thinking.

Like the endless litany of "whys?" asked by the preschooler, the question "Do you ever think about what you don't think about?" challenges those of us who work in mathematics education to explore, expand, and critique our work. Why do we teach division after multiplication? Why are students expected to know their multiplication facts by the time they go to middle school? Why do we still teach students to rationalize square roots? Why are mathematics and reading considered the two most important disciplines in school? Why is mathematical aptitude considered evidence of intelligence? Why does school mathematics emphasize continuous over discrete mathematics? Why do we teach 400-year-old algebra and calculus and 2500-year-old geometry? Why is mathematics viewed as so important and those who are mathematically competent showered with such esteem?

These questions, and many more, have puzzled me for years. Some answers may appear obvious but, when probed, they lose their clarity. Do we ever think about what we don't think about? Are there limits to what we can even think about? Why? Why mathematics? Why mathematics education? This chapter will explore those very questions in order to open up spaces for thinking about curriculum. I draw on ideas from Deleuze and use them, following the lead of Roy (2003), to develop perspectives that promise more dynamic curriculum futures.

Central to a Deleuzian cartography are interconnections of processes and relationships. Those interactions go further than relational ontology—they take issue with the very foundations of traditional philosophical categories of inquiry. Classic ontological and epistemological questions in the foundations of mathematics, and their search for essences, are replaced by surface or topological perspectives.

"Why mathematics?" will focus efforts to challenge the "in-between" or "nomadic" (Roy, 2003) spaces of knowing—to challenge us to think about that which we don't think about. By embracing these borderlands, the expectation is that our work in mathematics education will evolve creatively. I suggest that Deleuzian understandings can offer insights into the conundrum of mathematics as pure abstraction and extend our thinking about mathematics, mathematics education, and the curriculum.

THE SECRET OF THE CENTER: POSTSTRUCTURAL APPARITIONS

Postmodernists dissolve boundaries (Deleuze, 1968/1994) and shift centers. Twentieth century process and pragmatic philosophies, emerging from and capturing Hegel's "spirit" or the "geist" of Idealistic phenomenology, offer insights into educational futures through the decentering process. Dewey, anticipating the postmodern turn, confronted the question of the center in *The Quest for Certainty* (1929):

> The old center was mind knowing by means of an equipment of powers complete within itself, and merely exercised upon an antecedent external material equally complete in itself. The new center is indefinite interactions taking place within a course of nature which is not complete, but which is capable of direction to new and different results through the mediation of intentional operations ... There is a moving whole of interacting parts; a center emerges wherever there is effort to change them in a particular direction. (pp. 290–291)

The new center is not permanence, certainty, objectivity, and Truth but rather perpetual becomings, nonlinear relationship, organic process, and emerging meanings (Fleener, 2002). Rather than fearing the consequences of our own rejections of, and searching for, underlying realities or permanent structures, there is a freedom, a celebratory dance associated with recognizing that the center is itself in interaction with and emergent through our own quest for understanding. The joy, creativity, and re-enchantment with a world in perpetual becoming allow us to celebrate rather than bemoan the loss of certainty and structure: "We may gain the world by renouncing it, by passively losing self in the heart of what has neither form nor dimension" (Teilhord de Chardin, 1960, p. 21).

Postmodern perspectives, like Thoreau's castle in the air, are unfamiliar to those who are used to thinking in terms of foundations, hierarchical structures, underlying truths, certain knowledge, and given reality. It is difficult to escape our modernist language to describe postmodern perspectives, falling inevitably into "hierarchical-talk" or "foundational" analyses. We talk about "basics" or "building blocks" of knowledge in education, feeling with some confidence that, before students can learn, say, multiplication or fractions, that they must understand addition. The idea that the curriculum is somehow inherently structured and implies a sequence for learning activities contains a hidden assumption of hierarchy or foundations.

The hermeneutic, while historically connected to poststructuralism, creates the "inside-outside" separation associated with "thing" thinking, failing to escape the hidden ideas of foundations and hierarchies. From a poststructural perspective, however, "there's nothing transcendent, no Unity, Subject, Reason; there are only processes" (Deleuze, 1995, p. 145).

Early process thinkers, Bergson (1911) and Whitehead (1929/1978), in particular, had already shifted the focus from epistemological certainty, exploring the idea of "relationship" as emergence and the notion of "being" as complicated becomings. Challenging us to think about that which we do not ordinarily think about, twentieth century process philosophers created a new topology, a new playing field for philosophy, as inquiry, and with it, a different way of thinking—one that challenges the limits and assumptions of our modernist thinking. In attempting to accommodate theories of relativity, quantum indeterminacy, and evolution, 20th century process philosophers adopted a pragmatic approach to questions of underlying structures. Without necessarily rejecting realism or the assumptions of foundations, process philosophers, with their focus on interconnectedness, relationship, emergence, and self-reflective causality, have much to offer. Being, as process, implicates a complex web of relationship. Thus, at the center is the nexus—a complex of relationships and potentialities, "a dynamic process of creative advance that will never end" (Gragg, 1976, p. 16).

TOPOLOGICAL GEOMETRIES: A NEXUS

In common vernacular "nexus" has often been used synonymously with "center." This has become especially prevalent in advertising, business, and technology (see Buchanan, 2002). In the movie, *Star Trek Generations VII*, a "nexus" is a doorway to a timeless paradise and this interpretation provides another focal point but there the image is one of convergence of forces. More traditionally, "nexus" refers to sets of connections or links. With Latin roots in the verb "to bind," the word nexus first appeared in English in the 17th century to refer to "connection."[2]

Whitehead's process philosophy creates a topological field of interconnectedness, a nexus, based on experience. In his words, "A 'relation' between occasions is an eternal object illustrated in the complex of mutual prehensions by virtue of which those occasions constitute a nexus" (1929/1978, p. 194). Connecting experiences and constituting being as becomings, nexus, while relational, is more than just the connections or bindings of events into processes: it is "fundamental" reality. For process thinkers, the nexus is illusive and ethereal, fluid and dynamic, relational and interconnected. This idea shifts our perspective from the objects themselves to the processes and relationships of change. It is relational, organic and emergent. Just as the artist creates "negative" spaces to lend depth and dimension to the figures of focus in a painting, so, too, the nexus filters to the background, as "things" become the focus of our attention. But, what gives depth to our own lives (like the negative spaces in the painting) are not the accomplishments or products of our efforts, but the experiences themselves—the nexus of becomings as we engage life. As Whitehead (1929/1978) explains:

An actual entity is at once the product of the efficient past, and is also, in Spinoza's phrase, *causa sui*. Every philosophy recognizes, in some form or other, this factor of self-causation in what it takes to be ultimate actual fact. (Whitehead, 1929/1978, p. 228)

The notion of "causa sui" reemerged with process biology almost fifty years later. Bateson and Maturana recognized the "essence" of life not as some quality of "living" but as being the very nature of living things to engage in perpetual becomings, "causa sui," or self-making. Maturana invented the word "autopoiesis" to describe the process of self-creation, a poetry of self-making, that emphasizes that all of life is a nexus of becomings as we continually renew and re-create ourselves (Fleener, 2002).

Although I had read Whitehead and Dewey in my formal studies of epistemology of mathematics, it was not until I reread them as an educator that I realized they spoke very personally to me about my own teaching and my students' learning. Rediscovering Whitehead's (1929/1978) emphasis on relationship rather than the Newtonian primitives of entities in space and time significantly changed how I thought about schooling and my own teaching. *Causa sui* (self cause) and learning as a personal dynamic within a social context were ideas with which I intuitively felt comfortable. I changed from thinking about students (things) learning content (more things) in set periods of time (space-time) toward thinking about my students as complexes of relationships, autopoietic and dynamic. A new perspective of teaching emerged. Developing *relationships* with each other, with me, and with mathematics, as the focal point of my teaching, created a different kind of ethic—one that placed value on emergence and connections. Focusing on relationship, it seemed important to me to provide opportunities for my students to "fall in love" with mathematical ideas. *Ideas* themselves took on a very different role as well. In an ethic of relationship, ideas started to look like complexes of special kinds of relationship, and social networks of mathematical ideas now occupied the "in-between" spaces of interconnectedness, not "things" inside someone's head. The classroom became a conduit for the energy of those "in-between" spaces. Seeing connections was no longer an imposition of existing structures, miraculously "internalized" by my students, but was an exciting process of relational emergence. What inspired my teaching was helping students develop relationships with their own mathematical ideas and interests, helping them develop relationships with the mathematical conversations in the classroom, and helping them develop relationships with the discourse community of mathematics.

EXPLORING A DELEUZIAN CARTOGRAPHY

Poststructural approaches take the seeds of process perspectives and provide a theoretical framework that challenges the pragmatic silence (or indifference) to realism. Using process perspectives as a lever, poststructural perspectives have emerged and evolved from dialectical/phenomenological, existential and semiotic approaches to meaning. Jameson (1999) refers to these latter perspectives as "depth" models and they are to be distinguished from poststructuralism. Maintaining focal points and structures, albeit transitory, complex, and emergent, these postmodern transitional views are replaced, in poststructural approaches, by "surface" or topological models.

The topology of nexus, as a landscape of relationships, is organic, no longer placing humans or intelligence or abstraction at the center. The topology of nexus understands the interconnections of processes to lie in the *in-between spaces* of a world in flux. This understanding is clearly a challenge to the realism of most pragmatic and modernist thinkers. The topology of the nexus, as the ghost of postmodern process, however, may not be sufficient for challenging the dominance of mathematics. Nor might it be sufficient for opening up and extending our ideas about mathematics education, particularly in times of increased accountability and scripted curricula. Like the separation of worlds of science and the humanities, as described by C.P. Snow in his 1959 book *The Two Cultures,* or the operational duality of mathematicians, professionally approaching mathematics as formalists while personally embracing the permanence of mathematics as Platonic Forms to give their work meaning (as described by Hersh in his 1997 book *What is Mathematics, Really?*) the shift in seeing students in terms of relationships and processes was not sufficient, alone, to extend my ideas about curriculum practice.

As Spretnak (1997) describes:

> Although various conceptualizations of a relational, interdependent understanding of reality have been put forth ... they generally stop far short of accepting a radical nonduality. (p. 425)

Spretnak argues for a "radical nonduality" implicating a "unitive dimension of being." In support, Roy (2003) suggests the "impossibility of dealing with difference from the perspective of unity" (p. 9) and chooses to explore the resonance that "breaks through identitarian ways of thinking" (p. 9). My attempt at breaking through identitarian ways of thinking probes the idea of "radical nonduality" through a Deleuzian mapping. In my mind, the Deleuzian approach can challenge our thinking about mathematics, mathematics education, and the curriculum.

MAPPING CURRICULUM USING A NOMADIC TOPOS

The use of topological metaphors is deliberate in my mapping of the uncharted territories of mathematics, mathematics education, and curriculum. Just as the mathematics of topology introduced radically new ways of looking at relationships and by doing so, introduced new kinds of problems, so too do the topologies of "the center" provide opportunities to problematize and examine "that which we don't think about" at the core of our understandings. Roy (2003) describes a Deleuzian mapping that utilizes a nomadic topos as a map "unlike any other map; it is at once map and territory. It is nonrepresentational, which is to say, it does not represent but makes connections and projects new lines of flight" (p. 80).

Extending Deleuze's poststructural rejection of transcendental ideals, categorical analyses, and foundational approaches, nomadic topos creatively opens up new ways of seeing.

> [W]e reach a plane of multiplicities, a "nomadic" terrain whose cartography is based on flight from "striated" or highly regulated spaces where life's endless flux is coerced into preexisting moles or "molar" formations. (Roy, 2003, p. 14)

According to Roy, molarization occurs when we lose sense of the multiplicity and fluidity of the nexus, where ideas become isolated from their origins and relationships, and where boundaries are artificially constructed. Limits on our thinking occur precisely because we focus on "things" and forget the nexus, which itself is in perpetual becoming. One approach to overcoming molarization is to explore difference. This is not a new approach. Critical theorists, for example, "tend to employ a dialectical framework of analysis, and therefore, the analyses tend to explain relations in terms of opposing forces" (Fendler quoted in Roy, 2003, p. 36). The only way to overcome thinking in terms of the inclusive/exclusive is to break habits of thinking in terms of either/or or the dualism of the dialectic. As Roy explains:

> [T]he image of ourselves and reality is displaced from a universal and transcendent plane onto a *differential* and immanent one where we begin to act and move with the productivity of difference. (Roy, 2003, p. 31)

In his study of pre-service teachers, Roy (2003) utilized five interrelated spatial perspectives to explore the "nomadic spaces" of becoming teachers: (1) Smoothness; (2) Multiplicity; (3) Rhizoidness; (4) In-Betweenness; and (5) Becoming. When taken together, the traits "open up lines of continuous variation of nomadic space that deterritorializes the categories and boundaries within which conventional approaches to curriculum operate" (Roy, 2003, p. 72). It is Roy's contention that teachers and schools:

> … would be better served if they functioned with a differential cartography, rather than an identitarian one, and learn the new language of the mapping of intensities and becoming that leads to new possibilities. (Roy, 2003, p. 33)

The five categories of analysis that Roy has identified will be useful now as we explore the question: Why mathematics? within the nomadic topos.

Smoothness

Mathematics lends order to our world. Ignoring the bumps and holes, mathematicians create smooth spaces, generalization, uniformity, conformity, and predictability. These aspects of mathematics both frustrate and appease us; they are both boring and a comfort. Teaching and learning mathematics emphasizes the smoothness of relations. Plane geometry celebrates the perfection of idealized forms. The calculus smoothes out the infinitesimal, creating infinite limits and convergence. Even elementary arithmetic offers security and comfort in the logic of "2 + 2 = 4."

But non-Euclidean and fractal geometries, mathematics of recursion, topology and graph theory, even number theory and the foundations of mathematics challenge the smoothness, the intuitiveness, the confidence of what we typically think of as mathematical certainty. These approaches allow us to see the world, mathematics, and ourselves in a different light. What would it be to problematize the smoothness of mathematics, to explore the imperfections and irregularities of relationship ignored by most of classical mathematics? What might we come to see when we look for and challenge the regularities? Beyond the question "Why does this work?" how might we challenge the very core of mathematics and its smoothness?

There is a long history of mathematical innovators who have challenged the mathematical status quo. From the simple question "Why?" and subsequent question "What if?" other ways of organizing our world, of interacting with our world, of doing mathematics become possible. "The pits and tangles are more than blemishes distorting the classic shapes of Euclidian geometry. They are often the keys to the essence of a thing" (Gleick, 1987, p. 94). Mandelbrot explored these essences as challenges when he invented fractal mathematics. John Nash puzzled over von Neumann's non-zero-sum game theory and dared to ask "what if?" As Nasar (1998) argues: "the gaps and flaws in von Neumann's theory were as alluring as the puzzling absence of ether through which light waves were supposed to travel was to the young Einstein" (p. 87).

In our mathematics and mathematics methods classes for elementary education majors at the university level, we disrupt the smoothness of mathematics by problematizing the familiar. When asked "why do we invert and multiple when dividing fractions?" for example, our pre-service teachers experience their own gaps in their understandings of mathematics. They also come to see the value of problematizing the familiar, in disrupting the smoothness of "right answers" for the pursuit of the "why."

By pursuing the bumps and irregularities, rather than ignoring them or "smoothing them out," introducing complexity, challenging status quo, and questioning assumptions, the smoothness of mathematics is disrupted. In that pursuit the discontinuities, are not experienced as "blemishes" to be ameliorated in order to re-achieve balance, but rather are perceived as inviting new ways of seeing,[3] to experience the world as dynamic, to reinvent mathematics. They challenge the smoothness of mathematics and, in so doing, challenge us all to experience the creativity and emergence of mathematical ideas. They allow students to see their world and themselves as multiplicitous.

Multiplicity

The idea of multiplicity addresses the "impossibility of dealing with difference from the perspective of a unity ... it is then that a resonance occurs that breaks through identitarian ways of thinking" (Roy, 2003, p. 9). Multiplicity allows us to "relocate difference within repetition, in order to ... release the positivity of difference" (p. 12). Engaging the positivity of difference is necessary in order to escape molarization, and with it the loss of nexus fluidity.

When we engage the positivity of difference, ideas become connected and artificial boundaries are challenged. In turn, challenging rules, habits, and dispositions extends borders and celebrates multiplicity. Curriculum work could begin with a questioning of those assumptions, understandings, and meanings we take for granted, underlying the mathematics we teach, and the way we teach it. Part of the challenge of the questioning is to identify what Deleuze and Guattari call "order-words." By this they mean "the relation of every word or every statement to implicit presuppositions" (Deleuze & Guattari, cited in Roy, 2003, p. 28). The words "mathematics" and "algebra," for example, are order-words when used to reify ideas within a social setting. Many middle school students cringe at the thought of taking "algebra," an unknown yet feared type of mathematics. For them, "algebra" has become an order-word, unchallenged and unexplored when accepted without questioning. Within wider social settings, innumeracy is worn, sometimes like a badge of courage, in statements like "I was never very good at math." The word "mathematics" becomes an order-word as soon as we begin to think of it as static, or as meaning the same thing to everyone.

Within our own teaching, we forget to "listen" to our students, expecting that what we mean by a word or idea is what our students mean when they use the same expressions. Cabral (this volume) also emphasizes the point. It is an aspect of practice that became an issue in one recent study with which I was involved. One student responded "one and a half" at least six times in a ten-minute period to questions about how much of a pizza each of four children would receive if it were shared fairly (Fleener, Carter, & Reeder, in press). One possible meaning of the child's explanation of "one and a half" might be something like "one piece—a fractional piece." The teacher, understandably, was focused on trying to get the students to solve the problem and reply correctly with "one-fourth" and she overlooked this unexpected and persistent response. Multiplicity of mathematical ideas can encourage mathematical conversations. Exploring what the child meant by her response might have helped the teacher understand her student's use of the terms and might have enabled her to help the student connect her answers to the discourse community of the classroom. By pursuing her persistent use of "one and a half," the teacher might have also engaged other students in exploring their ideas about fractions and fractional relations.

In Roy's (2003) analysis, becoming teachers themselves are able to challenge their preconceived ideas about being a teacher when they understand the multiplicity of teaching roles. Instructor, disciplinarian, facilitator, listener, follower, leader, knower, explorer, investigator and so on, are more than just metaphors for teaching but revelations of the complexity of teaching. This point is brought out by Brown, Jones, and Bibby's exploration (this volume). The authors engage multiplicity by exploring the nexus of these relationships within teaching. Employing the conjunctive rather than disjunctive nature of the complexity of their roles, teachers, especially beginning teachers, are less likely to become embroiled in the contradictions of these roles as their teaching contexts change (see also Nichols & Tobin, 2000).

Viewing the role of teacher as multiplicitous supports identity multiplicity. Understanding students as multiplicitous is important when we try to make sense of their differences in understandings, interests, and performances. Once we recognize the multiplicity, we need to avoid reification. Thus, while we may recognize the multiplicitous nature of our students, we should never assume we've "figured them out" or grasped an essence of whom they are, for the nexus of their being is in continual renewal and reinvention. The dance of teaching is ongoing for we not only have multiple partners in the dance, but we and our partners are also multiplicitous and evolving. The terrain of multiple multiplicities, the nexus of multiplicities, the nexus of "nexi" can be explored as the Rhizoid.

The Construction of Multiplicities: Rhizomes

Deleuze and Guattari (1987) used the term "rhizome" to describe the "construction of a terrain of multiplicities" (Roy, 2003, p. 47). Avoiding hierarchies or levels of connectivity, the complex, emergent, and interconnected characteristics of the rhizome capture the systematic, organic and topological natures of the nexus. As described by Roy (2003):

> The "rhizome" is a lateral proliferation of connections, like the spread of moss, the sudden branching off or joining up of different intensities, flows, and densities to form new assemblies that have not fixed form or outline. A contingent mass, the rhizome can be cut up in any way and still retains operational wholeness; therefore it is highly tenacious. The rhizome is also a tuber, and unlike ordinary roots, can sprout in any direction. (p. 75)

The mathematics curriculum, as rhizome, is capable of emergent meanderings, discontinuous starts and stops, and the ebbs and flows of relationship. As the mathematics curriculum is allowed to "spread" through the ongoing experiences of students in the social contexts of the classroom, the web of relationships ensures viability. For example, a class of second graders exploring integers (typically a middle school concept) as a natural part of their making sense of number combinations or problematizing what happens with repeated subtraction (see Cassel, 2002) would exemplify the possibilities of a dynamic curriculum (Fleener, 2002).

One teacher with whom I have worked described his enactment of problem centered learning (PCL) as challenging the borders, disrupting the status quo, and encouraging the rhizome. PCL, for him, provided an avenue for students to pursue the "why" questions instead of concerning themselves with producing the right answer or getting through one problem to get to the next problem. As he describes PCL in his classroom, "What drives me to PCL is two-fold. The fact that over years of teaching I have had too many students trying to, in frustration, learn rules. Not asking "why?" but rather "what's next?" Not thinking for themselves, but willingly submitting to just plain conforming to someone else's thoughts because they are the all-knowing teacher" (Arbuckle, Che, & Matney, 2003, p. 6).

Standardization of the curriculum, or curriculum separated into "logically" framed, consumable "units" creates a stilted, non-adaptive, nonemergent nexus. The viability of the rhizome is associated with its wanderings, its complexity, its emergence, and its nonlinearity. Understanding the mathematics curriculum as hierarchical is to ignore the rhizomic nature of mathematics. To predefine the curriculum is precisely to stunt its growth and limit the complexity of relationships.

For individuals as well, the rhizome is a function of complex wanderings and explorations. "Teaching beyond" is nurturing the rhizome as nexus. Any classroom teacher can describe times when the students "took over" and ideas were flowing like sparks of electricity. In such a situation, students and teachers pursue ideas and make connections, not according to plan, but in a complex conversation with ideas. My own mathematics students knew I was amenable to "tangents" so would often ask me the "why" questions that took us off into the hinterlands of mathematics. After one such tangent, exploring Cantor's calculus of infinity and the paradoxes of the infinitely small, infinitely large, and time travel, my students perfectly described the potential of the rhizome. "Why can't we study the history of mathematics instead of precalculus? This was a lot more interesting and I learned a lot more mathematics." In my classes, I tried to nurture the growth of the rhizome by allowing students these tangents and pursuing these "teachable moments."

Wittgenstein refers to teaching beyond as "teaching etcetera" (see Fleener, Carter, & Reeder, in press). Genova (1995) describes "teaching etcetera" as an aspect of the rhizome that allows for openness and recursiveness of exploration. The openness of "teaching etcetera" is described by one teacher as a concern about creating spaces for students to continue to pursue problems. This teacher never "tells" students the right answers or expects the class to come to consensus at the end of the day. As he described to his students during my classroom observations:

> This problem does cause a struggle. If we don't get there by the end of class today I would hope that you would continue to think about it ... I'm not concerned with your taking this problem home and having your parents help you with it. I'm trying to get you to think about this. I care what you are thinking about it.

This hope is actualized as he may return to a particular problem months later. Students are encouraged to rethink their solutions, thinking in new ways about the mathematics, and to relate problems to differing and expanding mathematical contexts. In his classroom, students come to expect the meanderings, the recursions, and the twists and turns as problems and ideas are revisited and explored in multiple perspectives.

There are examples to quell the concerns of those who feel mathematics has its own internal structure or the learning of mathematics requires systematic and "scientifically proven methods" of instruction. Although many teachers feel pressure to follow closely a prescribed curriculum, many examples of highly regarded teachers who understand and enact the rhizome, have been recorded. The teachers described above have been recognized as outstanding teachers. Numerous studies have been carried out in their classrooms. The meanderings of the rhizome create a dynamic curriculum (Fleener, 2002), one that celebrates the richness, rigor, recursive, and relational complexities of learning (Doll, 1993). An environment that encourages divergence, meanderings, fixations, passions, and pursuits supports the viability of the rhizome that is the dynamic curriculum.

In-Betweenness

The Rhizoid curriculum, the multiplicitous and fluid nexus of individuals and social interactions, all shift our focus to the in-between, the relational, and the dynamic. No longer focused on "things" the in-between spaces need to be seen with "soft eyes."

Epstein (1999) describes how one of his patients learned to see with "soft eyes" to overcome hurdles in her equestrian jumps. She was frustrated because the harder she tried to make the last jump in competition the more she messed up. She was focusing, as she made her previous jump, on the last hurdle as the end of her performance. Then, the more she became aware of how her attention to the hurdle was affecting her performance, the more apprehensive she became. Her instructors taught her to "look past" the hurdle by imagining an additional maneuver she would need to make. Seeing with "soft eyes" allowed her to look beyond the jump, to experience the ride in process rather than focusing on the end results. Soon she was "in" the ride, experiencing the moment, no longer focused on the end.

Seeing with "soft eyes" is a way of acknowledging the in-between. As described by Roy (2003):

> To take in-betweenness not as a passage to something more definite but to treat it seriously, as an open space within every process, we have to understand how the teacher can act from the middle, from the in-between spaces, neither unifying instruction nor offering discrete packets aimed at different individuals. (p. 76)

Seeing our interactions with our students in the mathematics classroom with "soft eyes," looking beyond standards, end-of-year tests, or daily learning objectives, we can enact with our students the Rhizoid curriculum, the experience of mathematical exploration and creation. Rather than looking for or expecting a clear focus on their learning, it is my suggestion that we trust the nexus, look beyond the ends or products as measures of accomplishment and create the in-between spaces of dynamic interaction and relational being.

The teacher's role in nurturing the in-between, is to engage students and the school community in the nexus. As co-journeyers, students, parents, and principals might begin by opening up, with the teacher, the in-between spaces for learning. Engaging the in-between, students build their own understandings, not as foundations, but as complex webs of meaning, the nexus of relationship in the abstract world of mathematics.

In *The Birth of Tragedy*, Nietzsche described the Dionysian life as being lived in the in-between spaces of society and culture. To nurture the Dionysian spirit is to engage difference while seeking uniformity, balancing freedom and oppression, conformity and creativity. One way of nurturing the in-between is to encourage the Dionysian spirit in classrooms by engaging the "conjunctive-and" that rejects forced choice between two opposites. Thus, rather than either open-ended problem solving or routine practice, the conjunctive-and challenges students with mathematics that provides experiences as both routine and non-routine, exploration and practice, creativity and conformity. I used to explore paradoxes of time travel with my students, for example, as an extension of their own interests, connected to mathematics and the in-between of routine (equations of time dialation) and non-routine (fancies of paradox). To live in the in-between is to nurture the Dionysian in our students.

Thus, rather than a means to an end, the in-between is where the students are engaged in process, where the Dionysian spirit and rejection of forced dichotomies are encouraged. Infinite becomings, employing the conjunctive-and, our focus becomes blurred, no longer concentrating on "the target" or being constrained by "the basics" but riding Einstein's speeding bullet as we approach the speed of light in a thought-experiment. Like Zeno's paradox of the arrow, if, on the one hand, we think of motion as traversing the distance from the archer to the target, then, on the other hand, at incremental moments, the arrow is infinitely at rest. How, Zeno challenged, is motion possible if comprising infinite at-rests? Similarly we are inviting paradox when we fail to engage the in-between *as* the process, focusing on goals and ends rather than becomings and engaging the Dionysian spirit.

Trusting the potential of the in-between, teachers are encouraged to nurture a "plane of consistency" (Roy, 2003, p. 76).

At any point in time, the class is thus a rhizome, stretching and contracting between different point intensities, never unifying, nor becoming disparate.

And the teacher's position is always in-between, dancing between the lines.
(p. 77)

The nexus, as the in-between spaces of a world in flux, is in dynamic process. The nexus of the social setting, like the nexus of individual complexities, engages the in-between, the infinite becomings. Mathematics classrooms that engage the nexus, that approach the curriculum as Rhizome, honor the in-between by opening up spaces for meanderings, connections and explorations. From our paradoxes of time travel, we may explore paradoxes of the infinitely small, the continuum of real numbers, the calculus of infinities, and Cantor's geometric proofs of multiple infinities. We may come to see the density of the number line, not as infinitely full, but with periods of emptiness, of Cantor dust, and thus of fractal dimension. The foundations of mathematics, as simple as $1 + 1 = 2$, become suspect as we explore the paradoxes of statements like: "The sentence in quotes is false"—recreating Godel numbering and the incompleteness of mathematics.

Valuing and trusting the in-between acknowledges that classrooms, students, and learning are open systems. And in turn, the dynamics of openness require the flux of the in-between in order for openness to be maintained. As Whitehead (1929/1978) described, all being is in process, and the process is one of becoming.

Becoming

Becoming is understood as a process and implicates openness. Becoming, in one sense of the word, is the path taken or path made in the walking, as Freire would say. The book *The Man Who Knew Infinity* (Kanigel, 1991) describes the Indian mathematician Ramanujan's journey as a story of "becoming." "I have not trodden through the conventional regular course which is followed in a University course, but I am striking out a new path for myself" was the beginning of Ramanujan's letter of introduction to G. H. Hardy in 1913. Becoming is the path in its making, not the path to be taken.

We are all in a perpetual state of becoming. To support becoming, is to support the environment that nurtures and trusts in the potential of openness. Studies in dynamical systems recognize the importance of openness for self-organization. Over-prescription, control and planning thwart the creative advance of being, adaptive abilities, and self-organizing potential of individuals. Self-organization, as key to learning (Doll, 1993), is supported in open environments where multiplicity and difference are sustained and nurtured rather than controlled. Being, as becoming, is adaptive when self-organizing potential is supported by an open environment. Understanding the dynamics of mathematics, of social systems like classrooms, and of individuals as emergent, becoming-in-process emphasizes sustainability of environments that maintain openness and exchange of energies. The interconnectedness of the topos as explored by Roy (2003) and described above cannot be "seen" without engaging the positivity of difference. As process, engaging difference also requires exploration of the negative spaces of the nexus.

NEGATIVE SPACES AND CONCLUDING COMMENTS

We cannot enact the nomadic topos any other way than to engage in the process. Poststructural perspectives lay no claim to underlying realities or truths but can only hold together, like the famous ancient structures of South America, through organizational dynamics, the being, in process. At the risk of reification, creating structures-in-the-making, as emergent process, offers hope for a mathematics curriculum collapsing under its own weight.

Exploring the wanderings of the recursive "why" allows us to create the negative spaces that support mathematics and mathematics learning without changing their "realities" like Schrodinger's Cat of the famous quantum experiment. The interrelated spatial perspectives of smoothness, multiplicity, rhizoidness, in-betweenness, and becoming allow us to explore the uncharted, the relational, the recursive, and the emergent while also being discontinuous with gaps, fissures, and in-between spaces, in order to create the topologies of the "nomadic spaces" of mathematics learning. Deterritorializing the boundaries and limitations of the standard mathematics curriculum, extending the limits of our thinking, engaging the "why" of mathematics and mathematics education, a Deleuzian topology offers nomadic comings and goings and different ways of seeing the world of mathematics education.

NOTES

1.   "Inherit the Wind" is a 1960 movie based on the transcripts of the 1925 "Scopes Monkey Trial" in the United States that tested the legality of teaching Darwin's "Origin of the Species" and theory of evolution rather than the biblical version of the creation. The two opposing attorneys, Clarence Darrow and William Jennings Bryan, came to symbolize in the public press of the time two incommensurate perspectives: the modern scientific and traditional religious perspectives. The Spencer Tracy character, depicting

Clarence Darrow, however, rejects the dichotomy of, and opposition between, religion and science.

2.   See definitions of "nexus" at www.spellingbee.com/cc03/Week12/archive.htm.

3.   Elsewhere, I have discussed the importance of "seeing-as" differently (see Fleener, 2002; Fleener et al., in press). Building on Wittgenstein's distinction between "seeing" as a world picture or representation of the world, and "seeing as" as necessary for changing our world views and, fundamentally, as a way of life, the language games approach of Wittgenstein and the linguistic turn of postmodernism shift the focus from "knowing" to "seeing," which, in turn, implicates dynamic co-creation and poststructural approaches.

REFERENCES

Arbuckle, W., Che, M., & Matney, G. (2003). *Exploring the ramifications of enacting the spirit of problem centered learning in a standardized environment.* Paper presented at the Fourth Annual Conference on the Curriculum and Pedagogy, Atlanta, GA.

Bergson, H. (1911). *Creative evolution.* Henry Holt and Company.

Buchanan, M. (2002). *Nexus: Small worlds and the groundbreaking science of networks.* New York: W.W. Norton & Company.

Cassel, D. (2002). *Synergistic argumentation in a problem-centered learning environment.* Unpublished doctoral dissertation, University of Oklahoma, Norman.

Deleuze, G. (1968/1994). *Difference and repetition* (Trans: P. Patton). New York: Columbia University Press.

Deleuze, G. (1995). *Negotiations.* New York: Columbia University Press.

Deleuze, G., & Guattari, F. (1987). *A thousand plateaus: Capitalism and schizophrenia.* (Trans: B. Massumi). Minneapolis: University of Minnesota Press.

Dewey, J. (1929). *The quest for certainty: A study of the relation of knowledge and action.* New York: Minton, Balch & Company.

Doll, W. (1993). *A post-modern perspective on curriculum.* New York: Teachers College Press.

Epstein, M. (1999). *Going to pieces without falling apart: A Buddhist perspective on wholeness.* New York: Broadway Books.

Fleener, M.J. (2002). *Curriculum dynamics: Recreating heart.* New York: Peter Lang.

Fleener, M.J., Carter, A., Reeder, S. (in press). Language games in the mathematics classroom: Learning a way of life. *Journal of Curriculum Studies.*

Genova, J. (1995). *Wittgenstein: A way of seeing.* New York: Routledge Press.

Gleick, J. (1987). *Chaos: Making a new science.* New York: Penguin Books.

Gragg, A. (1976). *Makers of the modern theological mind: Charles Hartshorne.* Waco, TX: Word Books, Publisher.

Hersh, R. (1997). *What is mathematics, really?* New York: Oxford University Press.

Jameson, F. (1999). *Postmodernism or, the cultural logic of late capitalism.* Durham, NC: Duke University Press.

Kanigel, R. (1991). *The man who knew infinity: A life of the genius Ramanujan.* New York: Charles Scribner's Sons.

Nasar, S. (1998). *A beautiful mind.* New York: Simon & Schuster.

Nichols, S.E., & Tobin, K. (2000). Discursive practice among teachers co-learning during field-based elementary science teacher preparation. *Action in Teacher Education, 22*(2), 45–55.

Reeder, S. (2002). *Emergent mathematics curriculum: A case study of two teachers.* Unpublished Doctoral Dissertation, University of Oklahoma, Norman.

Roy, K. (2003). *Teachers in nomadic spaces: Deleuze and the curriculum.* New York: Peter Lang.

Snow, C. P. (1993). *The two cultures.* Cambridge: Cambridge University Press (Original work published 1959).

Spretnak, C. (1997). Radical nonduality in ecofeminist philosophy. In K.J. Warren (Ed.), *Ecofeminism: Women, culture and nature* (pp. 425–436). Bloomington: Indiana University Press.

Teilhard de Chardin, P. (1960). *Le Milieu divin: An essay on the interior life.* New York: Harper & Row.

Whitehead, A.N. (1929/1978). *Process and reality: An essay in cosmology.* New York: The Free Press.

CHAPTER 12

WHAT CAN I SAY, AND WHAT CAN I DO?

Archaeology, Narrative and Assessment

Tony Cotton

ABSTRACT

This chapter explores how the thinking and practice of one mathematics education researcher has been transformed through attention to the perspective of postmodern thought and through the use of postmodern theoretical tools. Lyotard's work on narrative and the Foucauldian notion of archaeology are particularly important in this chapter which explores the question 'What can I say, and what can I do?' and tackles the further question 'what can we say about what we do and what does what we say, do?'

INTRODUCTION

A man walks into a primary school classroom to collect his seven-year-old son. He sees the child sitting in a corner reading a book. He has pulled the hood on his sweatshirt right over his head, cutting himself off from the rest

Mathematics Education Within the Postmodern, pages 219–237
Copyright © 2004 by Information Age Publishing
All rights of reproduction in any form reserved.

of the class. The man smiles, then frowns—he remembers doing exactly this himself when he was a seven-year-old because he felt distant from all the other children in school and hated the mundane tasks that were set. He asks himself if this reaction can be genetic, and feels sad that his son feels the same separation from school and his peers as he did. Later he talks to his son about this. His son tells him that he is very happy with school, enjoys the activities, likes his teacher and has lots of good friends.

This felt to me like a postmodern moment. I had made assumptions about the reasons for particular actions, based on sensible evidence. My son had contradicted this. Of course, I would probably have said the same thing if my father had asked me the same question when I was seven. It was only later I (re)interpreted my own history and then imposed this on my son. There are no conclusions to be drawn, there are no facts to hang on to, but the episode recurs and the fact that I am (re)constructing it for you now means that it is open to further interpretation.

The task the authors contributing to this book were set was to provide a clear expression of terms crucial to postmodern thought in a way which is accessible to a wide range of readers. Since the implications for mathematics educational practice are important, this chapter, like the others in the volume, explores how our thinking and acting traditions might be transformed toward this radically different perspective. Postmodern thought cannot be pinned down—that is the point of the opening anecdote. Previously I have suggested that the most important thing we can do as researchers and teachers is become aware of "what, what we do, does" (Hardy & Cotton, 2000, p. 277), and hence the title for this chapter. However there is an additional problem for the researcher—the question here is "what can we say about what we do and what does what we say, do?"

This chapter draws on the Foucauldian notion of archaeology to develop ways of working toward transparency in mathematics education research and development. Foucault describes archaeology in his statement on the cover of *The Archaeology of Knowledge* (Foucault, 1972) as:

> ... an attempt to describe discourses. Not books (in relation to their authors), not theories (with their structures and coherences), but those familiar yet enigmatic groups of statements that are known as medicine, political economy, and biology. I would like to show that these unities form a number of autonomous, but not independent, domains, governed by rules, but in perpetual transformation, anonymous and without a subject, but imbuing a great many individual works (Foucault, 1972, back cover)

Archaeology explores how "things said" come into being, how they are interpreted, transformed and articulated. Archaeology, Foucault says, is the analysis of the archive consisting of the domain of "things said." The domain under investigation in this chapter is assessment in mathematics.

The chapter opens with an exploration of Lyotard's use of the terms "narrative" and "meta-narrative." The difficulties concerning representation that Lyotard poses lead me to prefer to describe what I write as description. This is followed by a section that moves the idea of presenting the unpresentable toward knowing the unknowable. How does postmodern thought impact on our view of knowledge? A third section offers an archaeology of assessment in mathematics education. Such an archaeology defines "not the thoughts, representations, images, themes, preoccupations that are concealed or revealed in discourses; but those discourses themselves, as practices obeying certain rules" (Foucault, 1972, p.138). The aim of such an archaeology is to expose the ideology present within current practice and through this description offer a view of possible futures. To end the chapter, I offer an approach to writing, influenced by the postmodern thinking of Lyotard and Foucault. The description of my work exploring assessment in a mathematics classroom works at the question "what can I do?" The focus on the notion of "narrative" works at the question, "what can I say?" The reader's task is to work at the question "what does what the author says, do?"

THE UNPRESENTABLE: NARRATIVE AND DESCRIPTION

Lyotard (1984) defines the postmodern as that which attempts to represent the unpresentable—that which searches for new ways to present work, not as a way of making the work more palatable, or more accessible, or even more exciting, but simply because we need to reinforce the point that what we are trying to do is present the unpresentable.

> The postmodern would be that which, in the modern, puts forward the unpresentable in presentation itself; that which denies itself the solace of good forms, the consensus of a taste which would make it possible to share collectively the nostalgia for the unattainable; that which searches for new presentations, not in order to enjoy them but in order to impart a stronger sense of the unpresentable. (p. 81)

This statement offers a challenge to writers of research in education. I would argue that what we observe in the classroom, what we notice as a response to our actions is too complex to be presented. My opening vignette was an attempt to illustrate this. Yet as writers we inhabit a Foucauldian domain that controls, both consciously and unconsciously, the way we present our thoughts and actions. The challenge is to search for new presentations that give a clear sense of the *unpresentable* nature of educational research and to write illustrating the complexity of human interaction. But we do not want to reduce this interaction to a simplistic series of "tips for teachers," however carefully these tips are framed.

Lyotard's work is also pertinent to the reporting of educational development and design as he re-interprets the death of the grand narrative. The demise of the grand narrative has always been a sticking point for educators working with postmodern ideas. In *The Postmodern Explained to Children* (Lyotard, 1992) he explains his position: "This is not to suggest that there are no longer any credible narratives at all. By meta-narratives or grand narratives, I mean precisely narrations with a legitimating function. Their decline does not stop countless other stories (minor and not so minor) from continuing to weave the fabric of everyday life" (Lyotard, 1992, p. 31). Lyotard himself suggests that his earlier work, in particular, *The Postmodern Condition,* "exaggerated the importance to be given to the narrative genre" (Lyotard, 1992, p. 31). But how else can we write but through narrative? How else can I write for an academic text but through narrative? Perhaps a way out of the impasse is to heed Griffiths' (1995) advice:

> Descriptions of experience are always revisable. I am assuming that I can recount my own experience without claiming that I am simply describing something independent of the description. I start from this situation and this situated self. I can recount this experience as it feels to me now, with my present level of understanding. It is quite possible that as I continue to think and theorize and observe, that I will understand more and my situated-self-understanding will change accordingly. (p. 14)

So perhaps I will move away from the use of "narrative" to describe this chapter. I consider it a *description* of my experience, from my current point of view.

Lyotard does not attack the use of research to explore (describe) the present situation and explore (describe) future possibilities. His attack is on the use of such research to present fixed views of the way things are or the way they should be. He notes the modernist focus on progress in politics and in education to emancipate the human race from "despotism, ignorance, barbarism and poverty" (Lyotard, 1992, p. 110). That focus, he concludes, has led humanity down blind alleys and has resulted in new tyrannies. Lyotard's work confronts the so-called crisis of representation. This affects the way we read and write text and influences the way we view knowledge. The way we view knowledge underpins the way we do our work in mathematics education. In the next section I explore the postmodern position on knowledge.

FICTIONS OPERATING AS TRUTHS

Knowledge in the postmodern age might be seen as the act of knowing the unknowable. The view I take is one that acknowledges the social construction of knowledge: we construct knowledge based on previous experiences and depending upon our present needs and the social situations in which we find ourselves. That view problematizes certainty. Walkerdine (1994) meticulously points out the dangers of searching for an ordered universe—a universe in which we can view everything with certainty. The enlightenment project is a dangerous one, she suggests. It leaves too many casualties in its wake. I have summarized the dangers she has described as follows:

Formal state education in the UK has never been organized for empowerment and liberation of oppressed groups, but rather functions to produce appropriate "subjects" for the social order of the time. The process by which this takes place is through producing "theories" about the nature of these "subjects," and the process of education, and pathologizing as deviant all of those who differ from this false norm. So we produce a theory that suggests that children learn best through early experiences of exploration, preferably located within a social situation with which children are familiar. They then need to use concrete objects to represent these situations. Eventually they are able to move into an abstract world. This theory becomes the way in which school curricula are planned and organized to the extent that there are no opportunities to learn in any other way. Pupils who do not learn according to this theory are deemed to have special needs and are offered remedial help until they learn how to learn correctly according to the dominant theoretical position. The assessments we design to measure children's learning are also planned and organized to support the dominant theory. These assessment processes support teachers in pathologizing those learners who do not fit the imposed norm.

Through such projects we produce what Foucault has described as "fictions which operate as truths." Ideas become embedded in the ways in which we work and the ways in which institutions function. For example, many teachers of mathematics with whom I work believe that setting/streaming/grouping pupils by "ability" improves "standards" in schools. In Nottinghamshire over the last ten years Government money has been available for primary schools to place pupils in "target groups'—groups organized according to test results. These pupils are then taught mathematics in these target groups. Pre- and post-tests appear to show that target-group pupils have improved their scores on tests at the end of the year, and this improvement is of course attributed to the "setting" of pupils in target groups. However, even if this is "true," even if we ignore the problematics of defining "ability" and "standards," even if we assume that the tests can measure competence accurately, there may be many reasons outside of the "target grouping" that have led to the changes. It may be to do with the increased resources available to the schools within the project; it may be to do with the positive attitudes of the teachers involved in the project; or it might have happened without target grouping (there are no control groups). However, the common sense view that setting pupils by ability in mathematics raises standards is heightened. In this way we manage to produce exactly what the "fictions" describe and the circle becomes almost impossible to break. In his work *Truth and Power in Power/Knowledge* Foucault (1980) describes how this circularity of truth is formed: "truth is linked in a circular relation with systems of power which produce and sustain it, and to effects of power which it induces and which extend it" (p. 133).

MATHEMATICS EDUCATION DEVELOPMENT IN A POSTMODERN WORLD

In the face of the postmodern understanding of knowledge production, what is the role of an educational researcher or developer? Michael Apple in the introduction to *Getting Smart* by Patti Lather suggests:

> We must shift the role of critical intellectuals from being universalizing spokespersons to acting as cultural workers whose task is to take away the barriers that prevent people for speaking for themselves. (Apple, 1991, p. ix)

Such critical intellectuals are clearly implicated in the production of knowledge. However their role is not to produce work that can be generalized to offer prescriptions for those who read the work. This is not an uncontroversial stance. Much educational research and development in recent years strive for the holy grail of "good practice." Elliott (1991) describes the way in which such generalizations in educational theory, "place teachers in what Ronald Laing called the 'double-bind situation' where a person is blamed for doing certain things, but given no indication of what (s)he could have done in the circumstances to have avoided blame. Generalizations ... tend to ignore the contingencies operating in particular practical settings, and provide no indication of how progress in realizing the ideal might be accomplished *in situ.*" (p. 47). Following from this, Apple's point about the role of a "critical intellectual" could be interpreted as providing readers and colleagues engaged in practice, an opportunity to interrogate how a particular piece of educational development constructs knowledge—how in itself the work is an historical and social construction.

Such knowledge can be defined as empowering. Lather describes empowerment as "analyzing ideas about the causes of powerlessness, recognizing systematic oppressive forces, and acting both individually and collectively to change the conditions of our lives. In such a view, empowerment is a process one undertakes for oneself; it is not something done 'to' or 'for' someone" (p. 4). Harvey (1990) supports the view of knowledge offered here, saying, "for critical methodologists, knowledge is a process of moving towards an understanding of the world and of the knowledge which structures our perceptions of that world" (p. 4). A critical emancipatory view such as Harvey's also offers readers of research an empowering approach to knowledge. My aim, as a "critical intellectual," is to make the text, for you as reader, critical and questioning.

ASSESSMENT IN MATHEMATICS EDUCATION

To tighten the focus on mathematics education research and development the next section offers an "archaeology" of assessment processes within mathematics education in primary schools in England. Foucault (1972) stresses that an archaeology is not a search for "preoccupations that are concealed or revealed in discourses" (p. 138) but a set of "discourses and practices obeying certain rules" (p. 138). The archaeologist must show how the set of rules governing the discourse is "irreducible to any other" (p. 139). For Foucault archaeology is "a regulated transformation of what has already been written … it is the systematic description of a discourse object" (p.140).

One of the purposes of archaeology is to uncover why things are the way they are: What are the rules that have led to the system we find ourselves in? Asking these questions allows those who want to make a difference to circumvent the obstacles in the way of change. Archaeology begins with context setting. In the late 1990s in the United Kingdom (UK), national assessment, particularly in mathematics education, attained a new prominence in terms of the measurement of pupil achievement, school achievement, and teacher performance. Schools and pupils were ranked, performance targets were set and sanctions applied to those schools and teachers who did not meet these targets. Although for many teachers in the UK the discourse of assessment, including the overtly discriminatory use of the results of such assessments, appeared to be new, formal examination systems have been in place for at least 2,220 years. A system of state written examinations existed in China around 200 BC. According to Griffiths and Howson (1974) it was devised in order to select government officials and acted as a way of:

- Screening (to select for a future role).
- Providing incentive.

- Stabilizing membership of the power hierarchy.
- Maintaining the tradition of scholarship.

Assessment in mathematics discriminates in each of these ways. If we accept that assessment will follow these rules and meet these purposes, it becomes clear that the ways in which such assessment is implemented plays a vital role in developing a mathematics education for social justice.

Assessment and Social Justice in the United Kingdom

It has been suggested that the new market environment in which education has been placed by the Conservative administrations of the 1980s, and which has been embraced by the New Labour administrations, has led to a values drift (Ball, Bowe, & Gewirtz, 1995, p. 150). In the new environment in which schools compete against each other for pupils and for the state funding that their enrolments guarantee, "morality itself is commodified: values are now treated as products to be packaged, marketed and sold" (Ball et al., 1995, p. 145). Not only do schools compete to move up the "league tables" through increasing pupils' examination success; they also compete for white liberal as well as minority ethnic parents through advertising their anti-racist and multi-cultural policies. This values drift can be seen in the move from an emphasis on student need to an emphasis on measurable student performance, particularly in the areas of Mathematics, English and Science. Similarly, there is a detectable drift from developing curricula for all ability groups toward the new common sense of grouping pupils by "ability." This ideological shift, together with the implications it has for assessment procedures, can be seen as a move away from comprehensive values, which embraced ideas of equity and justice toward the market values of the New Right (and New Labour).

This values drift has been accompanied by the introduction of a whole range of new high stakes tests. National tests at ages 7, 11, and 14 are used to assess pupils, their teachers and their schools and have had a huge impact on the education of the pupils within schools in the United Kingdom. Gipps (1994) reports on a review of research on the effects of high-stakes testing. I have noted key themes from the review:

- When test results are given high stakes by political pressure and media attention, scores can become inflated, thus giving a false impression of student achievement.
- High stakes tests narrow the curriculum. Tested content is taught to the exclusion of non-tested content.
- High stakes testing misdirects instruction, even for the basic skills.

- The kind of drill and practice instruction that tests reinforce is based on outmoded learning theory. Rather than improve learning, it actually denies students opportunities to develop thinking and problem solving skills.
- Because of the pressure on test scores, concepts more difficult to teach children are rejected by the system.
- The dictates of externally mandated tests reduce both the professional knowledge and status of teachers.

Gipps also points out how such a model of assessment, based as it is on individual competition, shapes the realities of many pupils in our classrooms. The Governmental response to discrimination in terms of educational opportunities and access to educational resources which becomes clear at times of high governmental control over assessment systems has been to encourage schools to "improve." Improvement, in this instance, is measured against the government definition of "effectiveness." There is an attempt to define what makes a "good teacher" in as simplistic a way as possible, and to encourage all teachers to take on the characteristics of these "good teachers." As a consequence it is possible to pathologize those pupils who do not succeed in the external examinations. If teachers endeavor to be "good teachers" following the curriculum, and if schools improve themselves, then the state absolves itself from blame for those who still fail within the system. They have only themselves to blame.

Assessment in Europe

In May 1992 a colloquium, "Differential Performance in Assessment at the end of Compulsory Schooling" was held at the University of Birmingham. *Who Counts? Assessing Mathematics in Europe* (Burton, 1994) was published from the colloquium and offered a view of the ways in which government and state ideology directly influence assessment systems in schools. But it also offered possibilities for change. The ten countries represented were Denmark, England, France, Germany, Ireland, The Netherlands, Norway, Portugal, Spain and Sweden. The representatives viewed both the purposes and practices of assessment in different ways. Denmark, Norway and Spain spokespersons adopted both philosophies and practices that supported a relativist approach to mathematics and mathematics learning. The other seven countries, to a greater or lesser extent, have adopted an absolutist curriculum driven by a national assessment base.

In Denmark end-of-grade exams were internally assessed by the teacher. Pupils could expect to have the same teacher throughout their school career. The final exams, signaling the end of school education, were optional and prepared centrally by a team of practicing teachers. Similarly, in Norway all assessment was school based and pupils were included in the process. The government view was that many different forms of assessment should be used across grades 1–6 with formal, but still school based, assessment in grades 7–9. Spain had a centralized curriculum but no external tests other than university set entrance examinations. Again the government line was that assessment in schools should be continuous; assessment was seen as part of teaching and learning—to be utilized for diagnostic purposes rather than for the purpose of ranking pupils.

Portugal, the Netherlands, and Sweden, although they all had centralized curricula and hierarchical school organization to some extent, demonstrated flexibility in their practices of mathematics assessment. In particular, the Netherlands had experimented with innovative, learner-centered approaches to assessment, designed to "challenge learning rather than maintain mathematical mysticism" (Burton, 1994, p. 9).

Regulating the Discourse

One way in which archaeology can show that a particular set of rules is regulating a discourse is by offering alternative criteria, and showing how things could be different if these were regulating the discourse. Within an alternative discourse an overriding principle might consider assessment as a dynamic force that sees student learning both as a part of the assessment and as a result of the assessment. In the UK, assessment developers have established five principles from which assessment models might be built. Assessment should:

- Improve learning.
- Test what is known rather than uncover what is not.
- Focus on the process of the curriculum objectives, not their products.
- Evaluate the quality of the test in ways other than objective scoring.
- Fit into the usual school practice.

The principles offer an alternative foundation upon which the discourse of assessment might operate. They are not the regulators that have driven the development of assessment in England—the system would have been different if they had been. Wiggins (1989) takes up the challenge to redefine assessment by suggesting criteria that should inform "authentic" assessment, built around four foci: structure and logistics, intellectual design features, grading and scoring standards, fairness and equity. Wiggins' description, summarized below, in my view offers a well developed, and clearly described set of criteria around which assessment might be structured:

A. Structure and Logistics
1. Assessment is public; it involves an audience, a panel and so on.
2. Assessment does not rely on unrealistic and arbitrary time constraints.
3. Assessment offers known, not secret questions and tasks.
4. Assessment consists of portfolios of work rather than one-off sessions.
5. Assessment requires collaboration with others as a part of the process.
6. Learners see assessments as worth practicing for, rehearsing and retaking.
7. Assessment and feedback to students are so central to the process of learning that school timetables, structures and policies are modified to support them.

B. Intellectual Design Features
1. Assessments are essential—not needlessly intrusive, arbitrary or contrived simply to grade a child for the purpose of ranking learners.
2. Assessments are "enabling"—constructed to point a student toward a more sophisticated use of the skill or knowledge.
3. Assessments are contextualized, complex intellectual challenges, not "atomized" tasks, corresponding to isolated outcomes.
4. Assessments involve the students' own research or use of knowledge, for which "content" is a means.
5. Assessments do not test recall or plug-in skills.
6. Assessments are representative challenges—designed to emphasize *depth* more than breadth.
7. Assessments are engaging and educational.
8. Assessments involve some ambiguous tasks or problems.

C. Grading and Scoring Standards
1. Assessments involve criteria that assess essentials, not easily counted errors.
2. Assessments are not graded on a curve but in reference to performance standards (criterion rather than norm referenced).
3. Assessments involve demystified criteria of success that appear to *students* as inherent in successful activity.
4. Self-assessment is a vital part of the system.
5. A multifaceted scoring system is used instead of one aggregate grade.

D. Fairness and Equity
1. Assessments identify (perhaps hidden) strengths.
2. Assessments strike a *constantly* examined balance between celebrating present achievement and prior experience.

3. Assessments minimize needless, unfair and demoralizing comparisons.
4. Assessments allow appropriate room for student learning styles, aptitudes and interests.
5. Assessments can be/should be attempted by all students, with the assessments scaffolded so all can access the process.

Such a system of assessment has been described as "high participation" (IPPR, 1990). A high participation system encourages students to take an active role in the education process, offers choice and allows negotiation of learning and thus encourages students to maintain a presence in the education system. In contrast, the present low participation system plays a disempowering function for the majority in school. Failing to "make the grade" results in educational access being denied.

In the next section of this chapter I describe the process of introducing a simple assessment system based on ideas of authenticity and high participation in a classroom in a primary school. This exemplifies an issue raised by Hardy in her chapter in this volume, concerning how the current discourses of mathematics pervade students' thinking.

MATHS AT WORK

To explore ways of writing influenced by postmodern thought, and drawing on ideas of Lyotard and Foucault as described earlier, I report on my introduction of an approach to assessment. The approach applies the criteria underpinning authentic and dynamic assessment, to a group of 10 and 11-year-old pupils in a primary school in England. National tests are routinely administered to pupils of this age range. I wanted to provide the children with an experience of long term planning and the opportunity to present mathematics through a sustained piece of writing. I also wanted to provide pupils with an opportunity to make choices and justify these choices as a support for their own learning. My hope was that by offering them the experience of working to a given set of assessment criteria, the hidden criteria of high stakes national testing would be counterbalanced.

The project was entitled "Maths at Work," and the activities were all taken from the mathematics program the class was following. The children's first task was to interview somebody they knew who used mathematics in their work or in their everyday life. They were asked to write a report based on this interview detailing their findings. The pupils then engaged in a range of activities exploring best buys in supermarkets, designing equipment to make the school "wheelchair accessible," exploring patterns in fabrics or carrying out data handling activities. The children could choose the activities, although I asked them to explain their choices as a part of the final report on the project. The children were expected to make individual appointments with me to discuss their personal action plans for the project.

My journal for April 18 states: *I managed to talk through all the action plans with the group. They have found it very difficult to see the purpose and to be realistic.* As I discussed the plans with pupils they were clearly expecting me to tell them when to do each activity, and to monitor their progress closely. The following extract from a pupil's journal illustrates both this process and the way in which the "project" became separated in pupils' minds from the "other" mathematics they were working on.

2/5/96 (Parmjit's journal)
We have been talking about the maths project that we are doing. I have started my two shape tessellation. I have started my patchwork as well. I need to do some HBJ because I have not done any because there were so many rehearsals for Billy.

9/5/96
I'm just a little bit nervous about the SATS [*Standard Assessment Tests*] because I think I might get a really bad score or level. I hope I get at least level 4 because I think it is my ability but I will be very happy if I get a level 5.

So even though the "project" Parmjit describes is based on the mathematics program followed by the school she has separated it in her mind as "project" work. She seems clear that the "expected" level is level 4. The teachers in the school had deliberately not emphasized the fact that the Government's expected level at the end of Key Stage 2 is level 4, although it was clear to many of the pupils.

An extended piece from Pardeep's journal shows the extent to which the expectations from the "Maths at Work" project produced different descriptions and measurements of success from the pupils.

26/4/96 (Pardeep's journal)
So far I have done 1 starter activity which was called "the Give Away." I liked this a lot because there was a lot of designing involved in it. You had to make a poster to show any product of your choice was a good buy. You could make it up or you could choose one that is out in the shop. I made one up which was called Filo.

Pardeep goes on to describe her poster in some detail—then she writes:

Last week we had a Maths practice SAT. I got 33/40 which was alright, but I wasn't happy. Yesterday we did a practice spelling SAT. I got in that 23/30 which is level 5, but I still wasn't happy. Also last week we had a Science SAT which I got 35/36/44 which I wasn't very pleased with either.

We see first one style of mathematics measured by the process and then unexpectedly the appearance of marks and levels. Later in her journal, Pardeep contrasts coming "highest in the class" in spelling with doing "the comic survey." She is "pleased" to be highest in the class and "excited" about her comic survey. One final journal entry from Sheetal illustrates this contrast further:

16/5/96 (Sheetal's journal)
In my HBJ project I thought I have done very well because I have done lots of work on it. When I came into school on Monday I felt a bit frightened because I didn't want to do the SATS but when I did the SATS it was half hard and half easy. I very didn't want to do the SATS but I had to.

It is perhaps unsurprising that as we approached the time for the self assessment process some of the pupils found the idea of assessing their own work difficult. I record a conversation with some of the pupils in my journal.

30/5/96 (Tony's journal)
... four girls in particular were very resistant to the idea that I would not be marking them [the projects].

In the end I had to assure these girls that I would mark their work as well as ask them to assess their own work. In this way they felt that they gained the external validation they desired for the project. This note comes at the beginning of a section in which I describe the process of "moderation" we worked through to enable the pupils to give themselves a mark for each section of their completed project. This moderation discussion supported the pupils in assigning marks to their work. For example, one criterion the children were marking against was their use of language.

Language

(High mark):   There are six steps which we needed to measure. We measured one step and multiplied it by 6 to find the total height. If the gradient was too steep a wheelchair would not be able to get up the slope.

(Low mark):   My pattern is very colorful with lots of shapes in it. They are all different sizes.

I outlined all of the criteria using examples of work from the pupils. After this process the group embarked on the task of marking their projects and assigning "effort" grades. This was on the pupils' insistence. It mimicked the process they were accustomed to with their own teachers. The marks they awarded themselves ranged from 50–93 (out of a possible 100) and the grades are shown below.

Grade	A	A–	B+	B	B–	C+	C	D	E
Number awarded	2	4	1	2	4	1	1	0	3

It was however the comments attached to their projects that proved illuminating in terms of the ways in which they had measured themselves.

One girl wrote:

> I am pleased with my "Making a Ramp," what makes a magazine popular and Egyptian triangles. I could improve my explanations for my work because I explained more. I think people would know a lot about how I understand my maths but not all the other things I do. (Grade A)

The comments show that the children were beginning to assess themselves to support their future learning rather than giving a summative mark or level. We see here justification for the grade along with aims for the future. Here is another assessment in the same mold:

> I was really pleased with my Interview because I learnt what my Dad does at work. I was also pleased with my "Make a Ramp" bit because I learnt how to measure steps as well as make them. I think my presentation was alright but something I would have worked on was my "Expensive Potato" because I didn't write any explanation. (I'd give myself an A-)

Not all the pupils gave themselves the highest grade. Sachin was honest about his mathematics during the half term.

> I did enjoy certain parts of the project especially the circles. I could have improved on it because I tried to get all the info ready including the interview by Friday after Tony's speech in the middle base. I admit I should get an ok grade but I have not done lots of explanation on the things I have done. (Grade C)

Sachin had left too much to the last minute and realized he had rushed things at the end. This was not an unusual situation for Sachin to find himself in—but this was the first time he had admitted that this was an area he needed to work on. It was also positive to see that he recognized his strengths. Finally Talvinder who missed many classes during the half term wrote: "I am not very pleased with my topic because I had missed some weeks" (Grade E).

This was more worrying as despite my support during the time I was in school and from support teachers, Talvinder remained very disappointed with the activities he had attempted. Clearly this is a place to start any review of the process in future. How could the process be improved to support Talvinder and the other two pupils who graded themselves E?

This section has explored the process of introducing assessment systems based on conflicting criteria to those within the dominant discourse. In terms of Foucault's idea about fictions functioning as truths, it illustrates how the dominant discourse of assessment stifles alternative ways of thinking. It shows how none of us can operate outside the discourse. It can also be read, however, as a narrative exploring the consequences of a disruption of the dominant discourse.

WHAT CAN I SAY?

Lyotard (1992) suggested that "reading is an exercise in listening" (p. 117). He also suggested "it is not up to us to *provide reality* but to invent allusions to what is conceivable but not presentable" (p. 24). With that advice I would like to engage in a piece of fiction and ask you to think about the writing and how you construct reality around it. The scene is a group of sixteen-year-old students in an English secondary classroom. The group is an "all ability" group—although this would not be a description the learners would recognize. For this group, in this school it has always been the case that all pupils are taught in tutor groups by teachers who remain the focus teacher for that group over their entire time in school. The school imported this idea from a school in Denmark with which the staff and pupils are regularly in contact. Although they rejected the idea of one teacher covering the whole curriculum with their tutor group, they have worked hard at constructing a timetable based around "focus" specialists.

On our visit the group is engaged in their end of secondary phase mathematics assessment. This takes place over a full week toward the end of the last term in school. It is a process they are used to because a similar process has taken place at some stage during each year of their time in school. Around the walls of the classroom are the assessment criteria, consisting both of process criteria and content criteria. Examples of both are:

- You must show that you can trial and evaluate a variety of approaches to solving a task and break a complex problem down into a series of tasks.
- You must present your ideas clearly showing you can use diagrams, symbols and graphs to explain your solutions.

- You need to show evidence of effective and efficient use of calculators and computers.
- You need to show your understanding of fractions and percentages.

The pupils are engaged in a series of extended tasks, which lead to a series of outcomes. Day 1 ends with a program of student presentations. Day 2 begins with pupils spending some quiet time evaluating their contribution to these presentations before they move into another group activity, in a different working group. This activity requires each group to produce a booklet summarizing the main points of specific topic within shape and space. During the third morning individuals work with one of these booklets looking for errors or inconsistencies and suggesting improvements. The group is used to working in this way as these booklets are often used as learning resources at all stages of their mathematics learning. During the afternoon the pupils work in a third group on a series of statistical data collection, representation and interpretation activities. They are also handed the individual test papers which they will sit on the final afternoon—this is to help them prepare and to support them in selecting the appropriate notes and materials they wish to bring into the paper. The data handling activity comes to an end during Day 4. Day 5 begins with the multiple choice test paper and this focuses, in particular, on what has become defined as "numeracy" in the UK. The final afternoon is taken up by each individual pupil preparing a portfolio gathered from the activities of the week. This portfolio is the evidence used to measure against the explicit criteria that the pupils have seen on the walls all week.

During the week a moderator from another local school has been present during two afternoons to offer advice to the class teacher and the pupils as to the judgments that are being made. All teachers are trained moderators to ensure that consistency is seen to be applied across schools. Both teacher and moderator can be seen encouraging pupils throughout the week to revisit parts of their work to give evidence of mathematical skills they have, but have not shown during a particular activity.

The teacher draws up a series of marks from the pupil portfolios corresponding to the criteria. There are five areas of mathematics that correspond to the discrete areas of the National Curriculum and each area is marked out of eight. This mark corresponds to the level the pupil has shown evidence of operating at during the week. These marks then go forward to an examination board at which the pupils are graded. Any discussion that takes place around borderline grades of the candidates also involves the students.

FINAL THOUGHTS

So what have I done? I have discussed Lyotard's ideas of narrative and meta-narrative, and the Foucauldian notion of archaeology. The objective was to offer an alternative perspective for assessment development away from the modernist approach in which the role of the developer/researcher is to offer explanations from a position of expertise. The view taken in this chapter considers the production of knowledge as a shared endeavor, accomplished through collaborative engagement with important ideas and questions. Such knowledge production is located in specific sociocultural settings and contexts, and such knowledge, if effective leads to thought and/or action.

I have tried to rise to the challenge of searching for new presentations that give a clear sense of the *unpresentable* nature of educational work. In that task I have attempted to demonstrate how the work I undertake is influenced by postmodern thought. I do not read the chapter as a postmodern critique of assessment and mathematics education but rather as a *description* of how postmodern thought has influenced the way I view assessment in mathematics education, and the way I write about what I see. The ideas I have described give a different focus to the epistemological basis for my work. In trying to reveal the unpresentable nature of all our work, I try to show why things are the way they are and how they might be different. This does not stop me from offering alternatives, but demands that I explore critically the nature of these alternatives.

REFERENCES

Apple, M. (1991). Series Editor's Introduction. In P. Lather, *Getting smart: Feminist research and pedagogy with/in the postmodern* (pp. vii-xi). New York: Routledge.

Ball, S., Bowe, R., & Gewirtz, S. (1995). *Markets, choice and equity in education.* Buckingham: Open University Press.

Burton, L. (Ed.) (1994). *Who counts? Assessing mathematics in Europe.* Stoke-on-Trent: Trentham Books.

Elliott, J. (1991). *Action research for educational change.* Buckingham: Oxford University Press.

Foucault, M. (1972). *The archaeology of knowledge and the discourse on language* (Trans: A. Sheridan). New York: Pantheon.

Foucault, M. (1980). *Power/knowledge: Selected interviews and other writings 1972–1977 Michel Foucault* (Trans: C. Gordon). New York: Pantheon Books.

Gipps, C.V. (1994). *Beyond testing: Towards a theory of educational assessment.* London: Falmer Press.

Griffiths, H.B., & Howson, A.G. (1974). *Mathematics, society and curricula.* Cambridge: Cambridge University Press.

Griffiths, M. (1995). *Feminisms and the self: The web of identity*. London: Routledge.

Hardy, T., & Cotton, T. (2000). Problematising culture and discourse for mathematics education research: Tools for research. In J.F. Matos & M. Santos (Eds.), *Proceedings of the second International Mathematics Education and Society Conference* (pp. 275–289). Lisbon: Centro de Investigação em Educação da Facultade de Ciências Universidade de Lisboa.

Harvey, L. (1990). *Critical social research*. London: Unwin Hyman.

Institute for Public Policy Research (IPPR) (1990). *A British "Baccalaureat": Ending the division between education and training*. London: IPPR.

Lyotard, J.F. (1984). *The postmodern condition: A report on knowledge*. Manchester: Manchester University Press.

Lyotard, J.F. (1992). *The postmodern explained to children. Correspondence: 1982–1985*. London: Turnaround.

Walkerdine, V. (1994). Reasoning in a post-modern age in mathematics. In P. Ernest (Ed.), *Mathematics, education and philosophy: An international perspective* (pp. 61–75). London: Falmer Press.

Wiggins, G. (1989). Teaching to the authentic test. *Educational Leadership, 49*(8), 38–50.

ABOUT THE CONTRIBUTORS

Tamara Bibby teaches mathematics education at King's College, London. She is interested in education policy and the ways in which policy impacts on primary teachers. Her research looks at these aspects and, in particular, the sense generalist teachers make of mathematics and its teaching. She has been involved in two research projects funded by the Nuffield Foundation: *Raising Attainment in Primary Numeracy* (1997), and *Mental Calculation: Implementations and Interpretations* (2003).

Tony Brown is presently Professor of Mathematics Education at the University of Waikato in New Zealand. He is on leave of absence from a similar position at the Manchester Metropolitan University. He has recently directed two projects for the United Kingdom government's Economic and Social Research Council, addressing issues of how student teachers learn to teach mathematics in primary schools. He has authored two books: *Mathematics Education and Language: Interpreting Hermeneutics and Post-structuralism*, published by Kluwer, and (with Liz Jones), *Action Research and Postmodernism: Congruence and Critique,* published by the Open University Press.

Leone Burton is the Series Editor for International Perspectives on Mathematics Education. She is a Visiting Professor at King's College, London and Professor Emerita at The University of Birmingham, in the United Kingdom. Her publications in mathematics education are widely known, the next book to come out being *Mathematicians as Enquirers: Learning about Learning Mathematics* to appear in 2004 with Kluwer Academic Publishers. This book results from her recent research focus on the links between epistemology and pedagogy within a social justice perspective.

Mathematics Education Within the Postmodern, pages 239–242
Copyright © 2004 by Information Age Publishing
All rights of reproduction in any form reserved.

Tânia Cabral obtained her doctoral degree, entitled *Contribution of Psychoanalysis to Mathematics Education,* from the University of São Paulo (USP) in 1998. Her interest in and study of psychoanalytical theory led to her membership of the Brazilian School of Psychoanalysis-SP. She bases her teaching and research on Lacanian ideas. Presently she holds a position at the recently created State University of Rio Grande do Sul (UERGS) where she teaches mathematics. As a researcher she investigates students' difficulties in learning calculus and interprets them in terms of Lacanian theory.

Tony Cotton has been working in education for 22 years. His work encompasses secondary school teaching, advisory and research. He currently works at Nottingham Trent University and is the program leader for undergraduate primary teacher training. His recent book *Improving Primary Schools, Improving Communities,* published by Trentham Press, is focused on issues of equality and fairness. His research and curriculum interests are in social justice and in this work he has formed close links with teacher education institutions in The Czech Republic, Denmark, Spain, Norway and Holland.

Paul Ernest is a Professor at Exeter University, UK, where he directs masters and doctoral programs in mathematics education. His research concerns fundamental questions about the nature of mathematics and how it relates to teaching, learning and society. He edits the *Philosophy of Mathematics Education Journal* at <http://www.ex.ac.uk/~PErnest/>. His books include *The Philosophy of Mathematics Education,* published by Falmer in 1991, and *Social Constructivism as a Philosophy of Mathematics,* published by SUNY Press in 1998. He is currently writing a book on the semiotics of mathematics education.

M. Jayne Fleener is Associate Dean for Research and Graduate Studies and Professor of Mathematics Education at the University of Oklahoma. Her teaching and research have been in the areas of philosophy, computer science, mathematics, mathematics education, educational futures and curriculum theory. She has written numerous publications and is author of the book *Curriculum Dynamics: Recreating Heart,* published in 2002 by Peter Lang. Her current research interests include exploring the complexity of classroom dynamics and poststructural possibilities for curriculum futures.

Tansy Hardy is a Senior Lecturer at the Mathematics Education Centre, Sheffield Hallam University. She brings her long experience in primary and secondary schools and in advisory work to her teaching. Her research interests are in teachers' professional development and in the nature of practitioner research. In particular, she is interested in the way teachers translate guidance on curriculum and pedagogy into patterns of practice in their own professional work, and the implications of that process for initial and in-service teacher education. She is an active member of the British Society for Research in the Learning of Mathematics, The Critical Maths Group and the International Group for the Psychology of Mathematics Education.

Liz Jones is a Senior Lecturer in Education at the Manchester Metropolitan University. She has a particular interest in Early Years and Educational Studies. Her appointment follows twenty years based in both mainstream and special primary schools. She has published widely and co-authored a book with Tony Brown, *Action Research and Postmodernism: Action Research and Postmodernism: Congruence and Critique*, published by the Open University Press.

Agnes Macmillan is a Senior Lecturer in the Faculty of Education at Charles Sturt University, New South Wales, Australia. Three decades of teaching in the early years of primary education led her a decade ago into tertiary education. Her doctoral studies at the University of Newcastle combined interests in language and literacy, sociological analytical frameworks and numeracy to examine what lies at the interface of informal and formal learning. She is the author of two books as well as articles and chapters in various early childhood and mathematics education publications.

Tamsin Meaney has worked as a teacher in many locations and this has contributed to her interest in the relationship between language and mathematics learning. She has worked with ESL students at a Technical and Further Education College in Sydney, with Aboriginal students at schools in remote communities in the Northern Territory of Australia, with teachers in the Republic of Kiribati, and with parents and teachers of a Māori immersion school in New Zealand. She has also authored junior secondary mathematics texts. Her present research involves analyzing the talk that Year 4 and Year 8 New Zealand students use in their mathematical problem solving discussions.

Jim Neyland is a Senior Lecturer in the School of Education at Victoria University in Wellington, New Zealand. He taught mathematics in high schools for 12 years, worked in pre-service teacher education for five years, held a senior position at the Victoria University Mathematics and Science Education Centre for two years, and spent two years working on curriculum development in mathematics. His masters thesis and PhD were in mathematical logic, and the ethics of mathematics education, respectively. He edited volumes 1 and 2 of *Mathematics Education: A Handbook for Teachers*, published by the National Council of Teachers of Mathematics.

Paola Valero is Assistant Professor at the Department of Education and Learning, Aalborg University, Denmark. She has worked in "una empresa docente," the research centre in mathematics education at the University of Los Andes, in Bogotá, Colombia. Her research is an interdisciplinary work exploring the political dimension of mathematics education. She is co-author of several books, including *La calidad de las matemáticas en secundaria. Actores y procesos en la institución educativa* ("una empresa docente", Bogotá, 1998). She has also co-edited several books, including *Researching the Socio-political Dimensions of Mathematics Education: Issues of Power in Theory and Methodology*, published by Kluwer.

Margaret Walshaw is a Senior Lecturer in the Department of Technology, Science and Mathematics Education at Massey University, New Zealand. She has taught mathematics to secondary school students, undergraduate students, and pre-service teachers. Her main interest is in making connections between mathematics education and social theories of the postmodern. Currently she is engaged as principal investigator in a Royal Society of New Zealand Marsden research project, exploring the development of numerate young women; and as principal co-investigator in a Ministry of Education project on teaching and learning numeracy.

AUTHOR INDEX

Mathematics Education Within the Postmodern, pages 243–247
Copyright © 2004 by Information Age Publishing

Steinbring, H., 28, 96
Stronach, I., 166, 167
Stubbs, M., 168

T

Taylor, C., 58
Teasley, S.D., 78
Teilhard de Chardin, P., 203
Thrupp, M., 56
Tiles, M., 16
Tobin, K., 210
Trivette, C.M., 189
        Troman, G., 56
Tymoczko, T., 16

V

Valero, P., 7, 36, 42, 43, 49, 106
Van Bendegem, J.P., 16, 25
Varela, F., 59, 60, 62
Vinner, S., 144
Vithal, R., 43, 52n.3, 106
Voigt, J., 40
Volosinov, V.N., 26
von Glasersfeld, E., 39, 123
Vygotsky, L., 20, 26, 39, 96, 124, 125

W

Walkerdine, V., 30, 106, 126, 168, 174, 223
Walshaw, M., ix, 122, 127, 130, 168
Watson, A., 125
Watson, N., 104
Waugh, P., 2
Weil, S., 70
Weinstein, J., 59, 62, 71n.3
Wells, G., 95
Wenger, E., 20, 78, 80, 90, 92, 95, 125
Whitehead, A.N., 203, 204, 205, 215
Whitenack, J., 40
Wiggins, G., 228
Wignell, P., 79
Wilson, D., 143, 144, 145
Wittgenstein, L., 15, 17, 21, 217n.3
Wolin, R., 127
Woods, P., 56
Wylie, C., 56

Z

Zevenbergen, R., 40, 106, 184
Zizek, S., 167

SUBJECT INDEX

Mathematics Education Within the Postmodern, pages 249–254
Copyright © 2004 by Information Age Publishing